Ernst Schering Research Foundation Workshop 22
Impact of Molecular Biology
and New Technical Developments
in Diagnostic Imaging

Springer
Berlin
Heidelberg
New York
Barcelona
Budapest
Hong Kong
London
Milan
Paris
Santa Clara
Singapore
Tokyo

Ernst Schering Research Foundation
Workshop 22

Impact of Molecular Biology and New Technical Developments in Diagnostic Imaging

W. Semmler, M. Schwaiger
Editors

With 54 Figures and 18 Tables

Springer

Series Editors: G. Stock and U.-F. Habenicht

ISBN-13: 978-3-642-64581-5 e-ISBN-13: 978-3-642-60844-5

DOI: 10.1007/978-3-642-60844-5

CIP data applied for

Die Deutsche Bibliothek – CIP-Einheitsaufnahme
Schering-Forschungsgesellschaft <Berlin>: Ernst Schering Research Foundation Workshop. - Berlin; Heidelberg; New York; Barcelona; Budapest; Hong Kong; London; Milan; Paris; Santa Clara; Singapore; Tokyo: Springer.
ISSN 0947-6075
22. Impact of molecular biology and new technical developments in diagnostic imaging. - 1997
Impact of molecular biology and new technical developments in diagnostic imaging / W. Semmler and M. Schwaiger ed. - Berlin; Heidelberg; New York; Barcelona; Budapest; Hong Kong; London; Milan; Paris; Santa Clara; Singapore; Tokyo: Springer, 1997
(Ernst Schering Research Foundation Workshop; 22)
ISBN-13: 978-3-642-64581-5

Softcover reprint of the hardcover 1st edition 1997

Typesetting: Data conversion by Springer-Verlag

SPIN: 10534132 13/3135–5 4 3 2 1 0 – Printed on acid-free paper

Preface

Currently, a disease can often be diagnosed by means of morphologic criteria using various imaging modalities. The clinical application of high resolution imaging by computed tomography (CT), magnetic resonance imaging (MRI), and ultrasound (US) has been shown to be extremely useful for diagnosis and staging of diseases in oncology, neurology, cardiology, and gynecology.

New developments focus on functional parameters pertaining to disease-specific medical imaging. Such new diagnostic strategies are possible using scintigraphic techniques and, more recently, MRI and US. In addition, new therapeutic concepts, including gene therapy, require specific tracers or contrast media for therapy monitoring, for example, enzyme activity and changes in receptor expression. For this purpose scientists conducting basic research, especially molecular biologists, and clinicians must collaborate in order to exploit the available interdisciplinary knowledge in the development of new imaging technologies incorporating the molecular signals of diseases.

These developments require a better understanding of molecular biology by clinicians and by researchers working in imaging technology. Similarly, biologists must understand the principles behind imaging techniques and recognize both the potentials and the limitations of the existing technology.

To provide a platform for the exchange of information, experts from both disciplines, molecular biology and imaging, were invited to this workshop to discuss the future development of "Molecular Imaging". The aims were as follows:

The Participants of the Workshop

- To provide an overview of existing approaches to functional and organ- and disease-specific imaging
- To discuss the limitations of existing imaging modalities and to predict the clinical demand for "Molecular Imaging"
- To introduce concepts of molecular targeting by discussing the pathophysiology of disease processes

This workshop succeeded well in initiating discussion among the invited speakers from the two fields, and the participants recognized the potentials as well as the difficulties in transferring ideas into workable and clinically adaptable solutions. In addition, the need for closer cooperation was well appreciated, as indicated by Dr. Wagner, who addressed the existing "gap " between biologic and clinical sciences.

The participants agreed that the primary and most important goal – a better understanding of each other's expertise – was achieved.

New ideas and cooperation were initiated which are likely to have a significant impact on the development of molecular imaging.

We are indebted to the participants for their contributions to the workshop and for their help in making it a success. Furthermore, we thank the Ernst Schering Researach Foundation for the generous support. We also thank Dr. Ursula Habenicht and Ms. Wanke for their excellent organization of the meeting.

Markus Schwaiger
Wolfhard Semmler

Table of Contents

1 Medical Imaging: Some Historical Aspects
E.J. Rummeny and P.E. Peters 1

2 Molecular Nuclear Medicine: From Genotype to Phenotype
H.N. Wagner, Jr. 15

3 Integrated Imaging of the Heart:
New Clinical and Experimental Approaches
E.A. Zerhouni 23

4 Principles and Clinical Applications
of Functional Brain Imaging
T.J. Brady, A.G. Sorensen, and B.R. Rosen 33

5 Molecular Probing and Imaging by Magnetic Resonance
Spectroscopy: Possibilities for Imaging Gene Therapy
B. Ross, S. Bluml, E.R. Danielsen, and K. Kanamori 43

6 Tumour Targeting Using Peptides, Monoclonal Antibodies
and Fragments
K.E. Britton 67

7 Future Challenges in the Use of Positron Emission Tomography
for In Vivo Molecular Imaging
T. Jones 83

8 Pharmacological Studies with Positron Emission Tomography
J.J. Frost 91

9 Use of Complex Carbohydrates as Targets and Ligands for Imaging
R. Tauber, S. Emig, B. Volz, C. Fieger, and C. Bührer . . . 103

10 Oligonucleotide Aptamers as Specific Targeting Devices in Diagnostics and Therapy
S. Wölfl, M. Kujau, A. Siebert, and M. Wolters 135

11 In Vivo Monitoring of Gene Therapy by Radiotracer Imaging
R.G. Blasberg and J.G. Tjuvajev 161

12 Intra- and Intercellular Recognition of Cell Signaling Using Oligonucleotides
M.K. Dewanjee 201

Subject Index 265

Previous Volumes Published in this Series 269

List of Editors and Contributors

Editors

M. Schwaiger
Technische Universität München, Klinikum rechts der Isar,
Ismaninger Str. 22, 81675 München, Germany

W. Semmler
Institut für Diagnostikforschung GmbH an der Freien Universität Berlin,
Spandauer Damm 130, 14050 Berlin-Charlottenburg, Germany

Contributors

R.G.Blasberg
Department of Neurology, Memorial Sloan-Kettering Cancer Center,
1275 New York Avenue, Room K923, New York, NY 10021-6007, USA

S. Bluml
Magnetic Resonance Spectroscopy Laboratory, Huntington Medical Research Institutes, 660 S. Fair Oaks Avenue, Pasadena, CA 91105, USA

T.J. Brady
MGH-NMR Center, Massachusetts General Hospital, Department of Radiology, 13th Street, Building 149, Charlestown, MA 02129-2060, USA

K.E. Britton
Department of Nuclear Medicine,Queen Elizabeth II Wing,
St. Bartholomew's Hospital, West Smithfield, London EC1A 9BE, UK

C. Bührer
Virchow-Klinikum, Kinderklinik-Neonatologie, Augustenburger Platz 1, 13353 Berlin, Germany

E.R. Danielsen
Huntington Medical Research Institutes, 660 S. Fair Oaks Avenue, Pasadena, CA 91105, USA

M.K Dewanjee
Department of Veterinary Biosciences, Radiology and Bioengineering, University of Illinois at Urban-Champaign, 3516 Veterinary Medicine Basic Sciences Building, 2001 South Lincoln Avenue, Urbana, IL 61801, USA

S. Emig
Virchow-Klinikum, Institut für Klinische Chemie/Biochemie, Augustenburger Platz 1, 13353 Berlin, Germany

C. Fieger
Virchow-Klinikum, Institut für Klinische Chemie/Biochemie, Augustenburger Platz 1, 13353 Berlin, Germany

J.J. Frost
Johns Hopkins University School of Medicine, B1-130, 600 North Wolfe Street, Baltimore, MD 21205, USA

T. Jones
Cyclotron Unit, MRC Clinical Sciences Centre, Hammersmith Hospital, Du Cane Road, London W12 ONN, UK

K. Kanamori
Huntington Medical Research Institutes, 660 S. Fair Oaks Avenue, Pasadena, CA 91105, USA

M. Kujau
Hans-Knöll-Institut für Naturstoff-Forschung, Beutenbergstr. 11, 07745 Jena, Germany

P.E. Peters†
Institut für Klinische Radiologie, Westfälische Wilhelms-Universität, Albert-Schweitzer-Str. 33, 48149 Münster, Germany

B.R. Rosen
MGH-NMR Center, Massachusetts General Hospital, Harvard Medical School, 13th Street, Building 149, Charlestown, MA 02129, USA

B.D. Ross
Magnetic Resonance Spectroscopy Unit, Huntington Medical Research Institutes, 660 South Fair Oaks Avenue, Pasadena, CA 91105, USA

E.J. Rummeny
Institut für Klinische Radiologie, Westfälische Wilhelms-Universität, Albert-Schweitzer-Str. 33, 48149 Münster, Germany

A. Siebert
Hans-Knöll-Institut für Naturstoff-Forschung, Beutenbergstr. 11, 07745 Jena, Germany

A.G. Sorensen
MGH-NMR Center, Massachusetts General Hospital, Harvard Medical School, 13th Street, Building 149, Charlestown, MA 02129, USA

R. Tauber
Virchow-Klinikum, Institut für Klinische Chemie und Biochemie, Augustenburger Platz 1, 13353 Berlin-Wedding, Germany

J.G. Tjuvajev
Department of Neurology, Memorial Sloan-Kettering Cancer Center, 1275 New York Avenue, Room K923, New York, NY 10021-6007, USA

B. Volz
Virchow-Klinikum, Institut für Klinische Chemie/Biochemie, Augustenburger Platz 1, 13353 Berlin, Germany

H.N. Wagner, Jr.
Divisions of Nuclear Medicine and Radiation Health Sciences, Johns Hopkins Medical Institutions, 615 North Wolfe Street, Room 2001, Baltimore, MD 21205-2179, USA

S. Wölfl
Hans-Knöll-Institut für Naturstoff-Forschung, Beutenbergstr. 11, 07745 Jena, Germany

M. Wolters
Institut für Diagnostikforschung GmbH, Freie Universität Berlin, Spandauer Damm 130, 14050 Berlin, Germany

E.A. Zerhouni
Johns Hopkins University, The Russell H. Morgan Department of Radiology and Radiological Science, 601 North Caroline Street, Room 4210, Baltimore, MD 21287-0842, USA

1 Medical Imaging: Some Historical Aspects

E.J. Rummeny and P.E. Peters †

1.1 Introduction . . . 1
1.2 W.C. Roentgen and His Discovery . . . 2
1.2.1 Biographical Aspects . . . 2
1.2.2 The Moment of the Discovery of a New Kind of Rays . . . 3
1.2.3 Spreading the News . . . 6
1.2.4 The Period Following the Discovery . . . 7
1.3 The Discovery and First Use of Contrast Media . . . 8
1.4 Nuclear Medicine: Some Historical Steps . . . 9
1.5 Development of Nuclear Magnetic Resonance: From the ^{1}H Peak to the Image . . . 11
References . . . 13

1.1 Introduction

During the last century many discoveries in the field of medical imaging were made leading to multiple modalities which are used now in day to day practice. During this workshop latest developments in research of medical imaging will be discussed. Lectures will be mostly focused on the development of nuclear medicine and MR-imaging. As an introduction to this field this article discribes some historical aspects of the discovery and the use of x-rays, nuclear medicine and nuclear magnetic resonance. It is based on cited articles and book chapters.

1.2 W.C. Roentgen and His Discovery

1.2.1 Biographical Aspects

Wilhelm Conrad Roentgen was born on 27 March 1895 in Lennep, a small German town, to Friedrich Conrad Roentgen and his wife Charlotte Constanze, née Frowein (Peters 1995; Glaser 1995). His father was a distinguished cloth manufacturer and merchant. The house he owned is still standing today and is part of the Roentgen Museum in Lennep, Germany.

Because of the fighting during the revolution W.C. Roentgen's parents sold the house in Lennep and moved to Apeldoorn (The Netherlands), the hometown of Charlotte Roentgen. W.C. Roentgen attended primary school there and was later sent to a private boarding school. Leaving home in 1861, he continued his studies in Utrecht, where a curious event took place which had considerable impact on his future life. For reasons which were never fully explained, he was expelled from school. According to Glaser's biography, one of Roentgen's classmates had drawn a caricature of the teacher on the fire screen. When the teacher arrived unexpectedly early he discovered the picture and became very angry about it. He asked Roentgen to name the "dirty" artist, which Roentgen refused to do. This was reason enough to suspend Roentgen from further lessions and to expel him from school. This seemed to put an early end to Roentgen's academic career.

However, in 1862, W.C. Roentgen enrolled at the Utrecht Technical School, a private institution which prepared students for entrance into technical high school during a 2-year course. Then, in 1865, he was enrolled for a short time as a visiting student in the Department of Philosophy at the University of Utrecht. In November 1865 he moved to Zürich, Switzerland, and attended the Polytechnic of Zürich (today Eidgenoössische Hochschule, ETH), which accepted students such as Roentgen who did not have a normal school certificate if they passed a demanding admisson test.

At the Polytechnic, Roentgen studied seriously and successsfully, and in 1868 he received his diploma in mechanical engineering. On 12 June, 1869, Roentgen obtained his Ph.D. degreee from the University of Zürich after his thesis, entitled "Studies on Gases", was accepted.

At the age of 24, W.C. Roentgen became assistant to Prof. August Kundt at the Department of Experimental Physics. Two years later, in 1870, Kundt accepted a Chair in Physics at the University of Würzburg, and Roentgen followed him there.

Roentgen's situation improved further in 1872, when Kundt was invited to the newly founded German University of Strassbourg (now in France, of course) and took his assistant with him once again. Roentgen was promoted to "Privat-Dozent" (private lecturer) and was appointed as a lecturer at Strassbourg University.

In 1886, he received an offer from the University of Jena, Germany, and, in 1888, another from the University of Utrecht, The Netherlands, both of which he declined. In 1888, however, Roentgen received an offer which he could not refuse: the University of Würzburg, where previously his academic career had been obstructed, now offered him the position of full professor and director of its highly reputed and well equipped Physics Institute. He accepted the appointment and moved back to Würzburg where, in 1894, he received the ultimate academic accolade in his election as Rector.

On 8 November 1895 he discovered a new kind of rays, a discovery which led the foundations for the development of our medical speciality and which led to a phenomenal expansion in research in the fields of physics, technology, and astronomy.

1.2.2 The Moment of the Discovery of a New Kind of Rays

The story of the discovery of "X-rays", as Roentgen called them, has been told in countless variations. This is due mainly to the fact that in his will Roentgen requested that on his death all his papers, including his laboratory books, be burned, unread. There ist thus much room for speculation and reatively little safe historical ground.

In his original communication Roentgen reported the experimental setup and his observation (Roentgen 1895):

> If one passes the discharges of a fairly large Ruhmkorff induction coil through a Hittorf vacuum tube, a sufficiently evacuated Lenard or Crookes tube, or a similar apparatus, and if one covers the tube with a rather closely fitting envelope of thin black cardboard, one observes in the completely

> darkened room that a piece of paper painted with barium platinocyanide lying near the apparatus glows brightly or becomes fluorescent with each discharge, regardless of whether the coated surface or the other side faces the discharge apparatus. The fluorescence is still visible at a distance of 2 m from the apparatus..... (Roentgen's original communication, translated by Otto Glasser).

Owing to the fact that W.C. Roentgen ordered that all his laboratory notes be destroyed, we must attempt to reconstruct the precise sequence of events from his own first communication and from the accounts of his friends and biographers.

It was late afternoon on Friday, 8 November 1895. W.C. Roentgen preferred to work alone in his laboratory, and on a Friday he could be fairly certain that nobody would disturb him. He had been conducting a series of experiments on cathode rays using a Lenard's tube – the one with the thin aluminium window – but, apparently, that night he decided to use a Hittorf or Crookes tube instead, without such a window.

After energizing the cathode ray tube he noted the well-known phenomenon of fluorescence of the tube's glass walls. He had darkened the room and was investigating the fluorescence of barium platinocyanide pasted onto a piece of cardboard. In order to avoid interference with fluorescence originating from the glass walls of the tube he wrapped the entire tube in black cardboard. After energizing the tube again, he made sure that the cardboard shielding was adequate. Despite efficient shielding, however, a faint glimmer was visible from the barium platinocyanide screen, which was more than 1 m away from the tube. He repeated the experiment several times and established beyond doubt that this phenomenon could not be due to ordinary light reflections, nor due to the cathode rays, because they did not travel so far through air.

Over the following 7 weeks Roentgen worked very hard in his laboratory investigating this phenomenon. It is said that he even ate his meals in the laboratory and had his bed moved there so that he could work without interruption.

Roentgen had already proved his capability as an experimental physicist, and the experimental set-up for the study of the "new light" was carefully planned. To document his observations he used photographic plates. On the evening of 22 December 1895, he asked his wife Bertha ot let him protograph her hand using the new rays. After a 15-min exposure the first radiograph of a human being was obtained, clearly

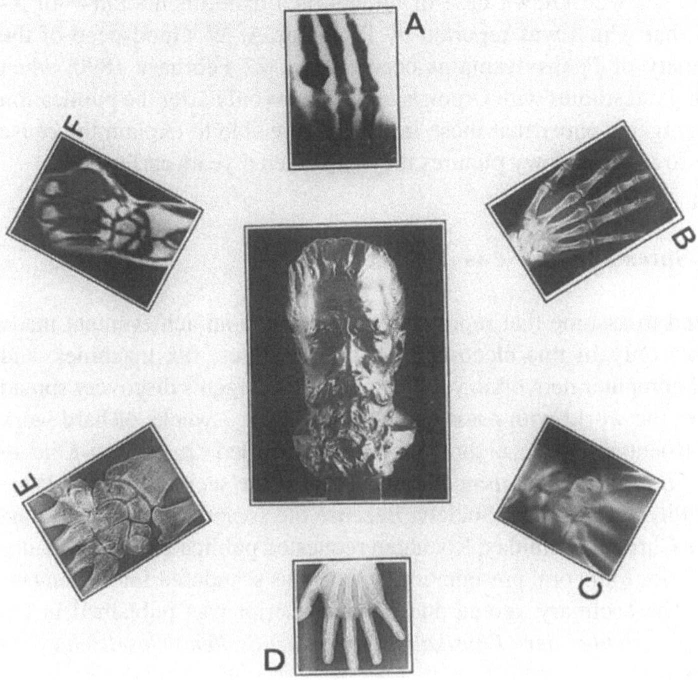

Fig. 1. Statue of W.C. Roentgen surrounded by images of the human hand obtained with different imaging modalities. From the first X-ray image to the magnetic resonance images. (Arranged and photographed by M. Hempel 1994); *A*, original X-ray of the hand from Roentgen's wife; *B*, conventional X-ray of the hand; *C*, conventional tomography of the wrist; *D*, CT scanogram; *E*, magnification radiogram of the wrist; *F*, MRI of the wrist

showing the bones of the hand and the two rings she was wearing. This date is the true birthday of radiology as a medical speciality (Fig. 1).

In retrospect it is apparent that many scientists unknowingly encountered X-rays in the course of their experiments. For example, Crookes complained to Ilford, his supplier of photographic plates, about fogged and blackened plates in unopened boxes. The company replaced the plates but observed that the damage must have occurred in the physicist's institute, since no other complaints had been received.

Another well-known case of unitentional experimentation with X-ray is that which was reported by Professor A. W. Goodspeed of the University of Pennsylvania as occurring on 22 February 1890, when carrying out studies with Crookes tubes. It was only after the publication of Roentgen's paper that those involved were able to explain the cause of the strange shadowy pictures they had taken 6 years earlier.

1.2.3 Spreading the News

We tend to assume that rapid communication is an achievement made possible only in this electronic age of statellites, fax machines, and global computer networks; yet the news of Roentgen's discovery spread all over the world with astonishing speed. After 7 weeks of hard work W.C. Roentgen wrote a short manuscript entitled *On a New Kind of Rays. First Communication* and handed it to the secretary of the Würzburg Physical Medical Society. Because the society did not meet during the Christmas holiday, Roentgen requested publication of the manuscript prior to its oral presentation, which was scheduled for 23 January 1896. The secretary agreed and the manuscript was publsihed in the *Sitzungsberichte der Physikalischen-medizinischen Gesellschaft in Würzburg*.

Three days later, on 1 January 1896, Roentgen had already obtained the offprints of the manuscript and he sent them, together with a few illustrative pictures, to his scientific colleagues and friends as a New Year's greeting. One of these offprints went to his old friend Prof. Exner, whom he had known since his college days in Zürich. Prof. Exner showed the manuscript and pictures to a group of physicists who were attending an informal scientific get-together at his home. One member of the group was Prof. Lecher from Prague who asked Exner if he might borrow the manuscript for a single day. Lecher's father was the editor of the Viennese newspaper *Die Presse*, and it was he who wrote the first article on the discovery, which appeared in the Sunday issue of that newspaper on 5 January 1896. Owing to haste in the preparation of the article the name of the Würzburg professor was misspelled as "Routgen." By the following evening, the news had been announced to the whole world by cable from London. The press notice read:

> The noise of war's alarm should not distract attention from the marvellous triumph of Science which is reported from Vienna. It is announced that Professor 'Routgen' (sic) of Würzburg has discovered a light which, for the purpose of photography, will penetrate wood, flesh and most other organic substances. The Professor has succeeded in photographing metal weights which were in a closed wooden case, also a man's hand, which shows only the bones, the flesh being invisible".

The *London Standard* printed the report on 7 January 1896, adding the following remarks:

> The Press assures its readers that there is no joke or humbug in the matter. It is a serious discovery by a serious German Professor.

While the first reports in New York were published on 8 January 1896, the first report by the local newspaper, *Würzburger Generalzeiger*, did not appear until 9 January 1896.

Thus, within 10 days of the submission of the manuscript to a publisher the news had spread to the entire world – before, even, Roentgen had presented his findings to the scientific society. On 23 January 1896, an English translation appeared in *Nature (London)*, and, 2 weeks later, in *Science* (USA).

On 13 January 1896, Roentgen presented his work to Emperor Wilhelm II in Berlin and, on 23 January 1896, he eventually gave his oral presentation to the Würzburg Physical Medical Society.

1.2.4 The Period Following the Discovery

W.C. Roentgen did not relish the sudden publicity he gained after the discovery of "a new kind of rays," which were named "Roentgen rays" following his oral presentation at Würzburg. He was awarded numerous honorary memberships, medals, and other decorations but virtually never accepted these accolades in person. He did travel to Stockholm, however, to accept the first Nobel Prize in Physics in 1901, but did not give an official speech even on that occasion. He was honoured by the Prince Regent of Barvaria with the Royal Order of Merit of the Bavarian Crown which carried with it personal nobility. Roentgen accepted the decoration but refused the status of nobility, a gesture which was most uncommon in those days in Germany. He turned down serval chances to make financial profit from his discovery. Mr. Levy, a represesentative of a well-known German company, was sent to Roentgen to negotiate a contract for the industiral exploitation of his current and future discoveries. Mr. Levy recalled Roentgen's answer:

> He declared, however, that according to the good tradition of German University professors he was of the opinion that his discovery and inventions belonged to humanity and that they should not in any way be hampered by patents, licences, contracts or be controlled by one group.

In 1900, W.C. Roentgen became Professor of Physics at the University of Munich and Director of the new Physics Institute. He retired in 1920 and died at the age of 78 on 10 February 1923 in Munich.

His life was devoted to research and his ideas still live. New technique using X-rays are still developing. Milestones in radiology which may be referred to in other chapters of this book include the development of computed tomography by Hounsfield and colleagues in 1968 and the introduction of digital substraction angiography by Mistretta et al. in 1977.

1.3 The Discovery and First Use of Contrast Media

The discovery of X-rays was immediately recognized as an event important to medical parctice. As early as 1896 Haschek and Lindenthal produced the first angiograms on amputated hands using Teichman's solution. In 1929 when thorium dioxide suspension was introduced, Egas Moritz, Professor of Neurology and Psychiatry in Lisbon, achieved spectacular success in carotid aorteriography (Grainger 1982).

The first glimmering of the successful visualization of the renal tract after oral or intravenous injection was published in 1923 by Osborne and coworkers from the Mayo Clinic in Rochester, USA. However, the major breakthrough in the development of X-ray contrast media was achvieved in 1925/1926, when Arthur Binz and his assistent Curt Reith synthesized organic iodine preparations at the Agrimentral College in Berlin. Although these compounds were developed for possible therapeutic action against syphilis and other bacterial infections, one series of pyridine compounds containing iodine was found to be excreted in substantial concentration by the kidney and the liver. It was the sodium salt of this product, later called "Uroselectan," with which Swick shortly afterwards produced the first successful and reliable intravenous urograms.

1928 and 1929 were truly momentous years in diagnostic radiology. At the same time as Swick was working with iodinated pyridine compounds, Egar Moniz and his surgical colleague Dos Santos were perfecting carotid arteriography and aortography. Also in 1929, a young German internist, Werner Forssmann, passed an uretic catheter from his left arm into his right artrium for angiocardiography. All contrast agents developed during the following years were ionic monomeric salts of tri-iodinated, fully substituted benzoic acids. The next major advance in the synthesis of X-ray contrast media was contributed in 1968 by Torsten Almen (Malmö, Sweden), who developed the first modern, nonionic contrast agent.

1.4 Nuclear Medicine: Some Historical Steps

Only shortly after the X-ray was discovered, Henri Becquerel made the discovery that mass is condensed energy. This was the birth of nuclear medicine (Wagner 1996). It began with physics, expanded into chemistry and instrumentation, and then greatly influenced various fields of medicine, such as endocrinology, cardiology, and the neurosciences. Important milestones in basic physics of nuclear medicine were set by Madame Marie Curie, who won the Nobel Prize in physics in 1903.

The invention of the tracer principle by Georg de Hevesy in 1912 illustrates the evolution of radiation research from the domain of physics to chemistry, and then in 1923 into the study of biochemistry. He was the first to elucidate what is perhaps the other important principle in biology – the principle of the dynamic state of body constituents. His first experiments showed how lead moved from the soil, through the body and then back into the soil.

The first use of radioactive tracers for diagnosis was performed at Harvard University in 1925 by Herman Blumgart. He injected solutions of radon in one arm vein and then measured how long it took to reach the other arm, calling the measurement the velocity of the circulation. Two years before, Ernest Lawrence began work on his cyclotron. The field of nuclear medicine is known for its use of radioactive tracers to measure in vivo biochemistry in experimental animals and in living human beings. One of the very earliest brain studies was performed in 1948 by George Moore, a neurosurgeon at the University of Minnesota, using di-iodofluorescein to localize brain tumors at surgery.

About the same time that Watson and Crick described the structure of DNA, Benedict Cassen and his colleagues invented the rectilinear scanner based on the use of scintillation detectors rather than Geiger-Mueller tubes.

In 1958 Hal Anger presented his first scintillation camera. What made the scintillation camera so valuable was ^{99m}Tc. Paul Harper and Catherine Lathrop realized the wonderful physical characteristics of ^{99m}Tc and began to produce a whole series of useful compounds. The scintillation camera made it possible to perform dynamic studies, which led to nuclear cardiology.

The first tracer used to examine myocardial perfusion was ^{43}K. It resulted in very poor images; therefore the motion of the heart was stopped by electrocardiographic "gating" of the scintillation camera. Investigators at the National Institutes of Health divided the cardiac cycle into 16 time frames instead of just end-systole and end-diastole, making ventriculography what it is today. Tomography was invented by David Kuhl in the early 1960s. He performed the first [^{18}F]FDG (fluorodeoxyglucose) studies on his Marc IV positron-emitting tomography (PET). Michael Phelps, Michel Ter-Pogossian, Edward Hoffman, and their colleagues then worked on the further development of PET.

In 1963, nuclear medicine expanded into pulmonology when the first lung scan was performed in a patient with massive pulmonary embolism. The use of lung scanning in the urokinase pulmonary embolism trial led to quantification in nuclear medicine by computers.

Almost 50 years ago, Otto Loewi discovered chemical neurotransmission, which earned him the Nobel Prize. Later, in monkey studies, Louis Sokoloff showed that closing one of the monkey's eyes resulted in decreased glucose utilization in the cerebral cortex. Wolfe and Kuhl discovered that which FDG many tumors are characterized by the avid accumulation of glucose – a major new advance in nuclear medicine.

PET and SPECT (single photon emission computerized tomography) have advanced together since 1982. Both are essential for the further development of the field. Nothing, perhaps, has brought them more closely together than the use of dual-headed cameras operating in the coincidence mode or so-called "SPECT scanning" using tracers such as FDG.

Today, nuclear medicine is being applied in oncology and genetics. In the development of this field each link in the chain of events has led to the subsequent link, and the chain is still expanding into new fields of medicine.

1.5 Development of Nuclear Magnetic Resonance: From the ^{1}H Peak to the Image

Nuclear magnetic resonance (NMR) has been the subject of intense research since 1936 (Mouring 1991). NMR was first reported in a solid in 1946 by two research teams: one led by Felix Bloch, the other by Edward Purcell. These two teams worked independently on this phenomenon and were unaware of the other's research. For their work on NMR, both received the Nobel Prize in physics in 1952.

NMR developed quickly as a scientific discipline. Bloch introduced the concepts of a "thermal" or "longitudinal" relaxation time (T1) and a "transversal" relaxation time (T2) in a seminal article he wrote in 1946. In that same article he derived the "Bloch equations," which are still the basis of all analyses of NMR experiments. He also provided the rationale for the use of those substances that are called contrast agents in magnetic resonance imaging.

The need to sustain the NMR signal over extended periods of time for the measurement of NMR phenomena spurred Erwin Hahn to the development of the spin echo in 1950. This technique was subsequently improved in 1954 by Herman Carr and Edward Purcell. The variation on their work, which was done by Saul Meiboom and David Gill in 1958, would one day produce one of the standard imaging pulse sequences – what we now call the Carr-Purcell-Meiboom-Gill spin-echo pulse sequence. The introduction in 1966 by Richard Ernst and Weston Anderson of Fourier transform techniques to the analysis of magnetic resonance phenomena marks the moment when the technical and analytic armamentarium necessary for nuclear magnetic resonance imaging became complete. Fourier transform NMR was employed in the study of molecular dynamics by Rober Vold in 1968 and by Raymond Freeman in 1969. The application to magnetic resonance imaging of their experimental technique would yield the second standard imaging pulse sequence, the inversion-recovery pulse sequence.

"NMR evidence for the existence of ordered water" in living cells was the research goal for many scientists during1965-1971. It emerged from the NMR spectra of water in systems as varied as relaxed and contracted frog skeletal muscle, partially dried yeast cells, decomposing fish muscle, oriented (with respect to the NMR spectrometer magnetic field) rabbit sciatic nerve, and rat skeletal muscle. It was only a matter of time before someone would try to see what, if any, differences appeared in the NMR "signature" of a tissue as a consequence of disease – and therein the tale of the origin of magnetic resonance imaging (MRI) can be found.

In 1971, Raymond Damadian reported that NMR could be used to discriminate between malignant tumors and normal tissue. Efforts to reproduce this work were quickly set in motion. In September 1971, Paul Lauterbur watched Leon Saryan as he compiled the NMR spectra of different rat tissues. In Lauterbur's own words "although there were clear differences .... there seemed no plausible reason for the differences... even normal tissues differed markedly among themselves in NMR relaxation times, and I wondered whether there might be some way to nonivasively map out such quantities within the body. The principle upon which a technique might be based, the encoding of spatial coordinates by known magnetic field shapes, occurred to me the same evening. Over the next several days, a general method ... to generate a true image ... became clear." The terse paper that he sent to *Nature* describing his methods was rejected because "it was not of sufficiently wide significance for inclusion in *Nature*." By writing a lengthy rejoinder to this judgment and rewriting his manuscript in a more exuberant style, he produced the result that he desired – the paper was accepted by *Nature* and published in the 16 March 1973, issue.

Lauterbur was able to publish his first image – a cross-sectional image of two vials of water – in 1973. Within a year of the appearance of that image, four methods of MR imaging had been demonstrated by other geographically disparate groups; the race to produce the first MR image of a human had begun.

Milestones in NMR include clinical application of MR spectroscopy, functional MRI, and the development of tissue specific MR contrast media.

References

Glasser O (1995) Wilhelm Conrad Röntgen und die Geschichte der Röntgenstrahlen. 2. Aufl. Springer Verlag, Berlin, Heidelberg

Grainger RG (1982) Intravascular contrast media – the past, the present and the future. Br J Radiol 5 (649):1

Mouring MR (1991) From Thales to Lauterbur, or from the lodestone to MR imgaging: magnetism and medicine. Radiology 180: 593

Peters PE (1995) W.C. Roentgen and the discovery of x-rays. In: Peterson H (ed) A global textbook of radiology. NICER Institue, Lund, Oslo. pp1–11

Wagner HN (1996) Nuclear medicine: 100 years in the making. J Nucl Med 37(10): 18N

2 Molecular Nuclear Medicine: From Genotype to Phenotype

H.N. Wagner, Jr.

2.1 Introduction 15
2.2 Homeostasis 17
2.3 Stress Tests 18
2.4 Multidrug Resistance 18
2.5 From Risk to Reality 19
2.6 Dementia Susceptibility Genes 20
2.7 Inherited Metabolic Disorders 21
2.8 Monitoring Gene Therapy 21
References 22

2.1 Introduction

No field of medicine is better able to translate the revolutionary advances in molecular biology and genetics into the care of patients than the field of molecular nuclear medicine. The human genome project has been a stimulus to the development of what might be called "genetic sciences," i.e., molecular genetics, human genetics, molecular nuclear medicine, and pharmacology.

The revolutionary advances in molecular biology and genetics demand a new approach to the definition of disease. Present criteria for defining most diseases are too imprecise and interfere with attempts to determine the relative roles of hereditary and environmental factors in diseases, such as Alzheimer's, Parkinson's, and other diseases. The

International Classification of Diseases employs a six-digit number to characterize a disease: the first three digits concern the anatomical location of the disease, while the second three digits concern etiology, i.e., whether the cause of the disease is infectious, traumatic, metabolic, etc.

The use of radioactive tracers to characterize disease processes is based on measurement of regional biochemistry and physiology, which makes it possible to examine the inheritance of both local and global disease. Global effects of mutations can often be studied by assay of enzymes or genes in circulating cells or body fluids, but mutations often have local rather than systemic effects. Pathogenesis (i.e., characterizing abnormal regional biochemical/physiological processes) is not the same as etiology (i.e., the designation of the cause or causes of a disease). I propose a new classification system that would be based on nine rather than six digits as in the present international system. The first three would concern location; the next three would be concerned with the altered biochemical or physiological processes; and the last three would concern etiology, which in many cases is unknown. The nine digits would characterize a "disorder," not a disease. A disease would be comprised of one or more specific disorders, the latter defined by global or regional biochemical or physiological processes.

The International Classification of Diseases, which uses only location and etiology, is usually insufficiently specific for the demands of genetic investigations. Often the cause of a disease is unknown or cannot be cast in simple, unqualified terms. For example, hypercholesterolemia may cause myocardial infarction, but other etiological factors are also involved.

To explore the inheritance of a disease, before the population is chosen and the data are collected, one needs a more specific characterization than is possible with the six-digit approach to diagnosis. Errors (misdiagnoses) are commonplace. For example, in the diagnosis of Alzheimer's disease, the use of clinical or psychological criteria results in misdiagnoses in 35% of patients, with the correct diagnosis being based on histopathological criteria in the same patients. Characterization of patients using radioactive tracers and computed-tomography studies was able to reduce this error rate to 3%.

Genomic classification of a patient is not sufficient either. While at times a single mutant gene can have a large effect, producing an abnormal protein, such as a clotting factor, most genetic disease is pleiotropic, i.e., the defective gene results in abnormalities of two or more organs. Diseases caused by a single gene anormality, such as cystic fibrosis or sickle cell anemia, are rare. Furthermore, disease such as cancer can result from several mutations and may be greatly influenced by environmental factors, such as smoking, so that the pathogenesis is complex and may manifest numerous regional biochemical abnormalities. Most cancers involve mutations in many genes.

Furthermore, a single gene may result in the expression of several proteins or peptides.

2.2 Homeostasis

In order to survive long and arduous pregnancies and subsequent life stresses, a human being relies on countless feedback mechanisms that have arisen in the process of evolution. Dead reckoning in biological processes cannot meet the needs of a complex, stressful life, any more than they can be used to land a space vehicle on the moon. Feedback mechanisms evolved because of their selective advantage.

The father of modern physiology, Claude Bernard, described the importance of biological feedback processes in maintaining the constancy of the "internal environment." i.e., the extracellular fluid of the body, a concept extended by Walter Cannon and called homeostasis. E.A. Murphy has pointed out that the individual processes going on in a complex biological consortium are often obscured because of the operation of these feedback mechanisms. An example is stability of arterial blood pressure, which is little affected by processes such as gravity unless the autonomic nervous system is impaired and orthostatic hypotension develops. Drugs such as pitressin are not observed to have a pressor effect in normal individuals, but result in a dramatic increase in blood pressure when compensatory mechanisms are lost in patients with orthostatic hypotension.

2.3 Stress Tests

Radioactive tracers make it possible to study the regional biochemical and physiological responses to perturbations that involve homeostatic processes. An example is activation of insulin secretion in response to a glucose infusion.

Some homeostatic processes are relatively easy to study. Processes in which phenotypical responses have been linked to corresponding genotypes include the following:

1. *p*-Glycoprotein expression in response to toxins
2. Glucose transport in the brain and heart
3. β-Oxidation of fatty acids in the heart

2.4 Multidrug Resistance

An important recent contribution of nuclear medicine to genetics research, which also has practical clinical usefulness, is the finding that technetium-99m-(^{99m}Tc) Sestamibi can be used as a chemical phenotypic marker of the expression of the human multidrug resistance gene MDR1. This gene leads to the overexpression of *p*-glycoprotein, an enzyme responsible for detoxification of certain foreign chemicals in the body.

In patients with osteosarcoma, the response to chemotherapy depends on the degree of expression of *p*-glycoprotein. Failure of patients with small cell lung cancer to respond to chemotherapy can be predicted by measuring the accumulation and release of ^{99m}Tc-Sestamibi from the lesions (Bom et al. 1996). When the levels are low, the response is poor, presumably because of the rapid excretion (detoxification) of the chemotherapeutic agent.

In a effort to improve the effectiveness of chemotherapy by modulating *p*-glycoprotein effects, Franssen and colleagues from Groningen University Hospital in the Netherlands administered PSC 833, a cyclosporin analogue, and found a reduction of ^{99m}Tc-Sestamibi efflux from the liver and tumors of patients (Franssen et al. 1966). Piwnica-Worms and colleagues from Washington University Medical School in St. Louis have developed a Gallium-67-labeled phenolate complex that will make it possible to characterize *p*-glycoprotein activity with the positron-emitting tracer gallium-68 (Sharma et al. 1996).

In patients with lymphoma, the accumulation of thallium-201 combined with poor retention of ^{99m}Tc-Sestamibi indicated that there would be a poor response to chemotherapy (Kapucu et al. 1996).

2.5 From Risk to Reality

Assessing a person's genetic risk of developing a specific disease will be a major task of medicine in the 21st century. All or nearly all of the human genome will have been sequenced early in the next millennium. Susceptibility genes for cancer include BRCA1 and BRCA2 for breast cancer, AT for breast cancer, and p53 for nearly half of all cancers. The detection of gene mutations will tell us the probability that a patient will develop an illness sometime in the future. What we really need to know is what each gene does and how it does it.

More and more clinics now provide DNA-based susceptibility testing for mutations in several such genes to help identify those women at exceptional risk for developing breast or ovarian cancer. Inherited breast cancer is thought to account for 5%–10% of all cases of breast cancer. Lifetime risk for BRCA2 mutation carriers is estimated to be 85%, and risk for ovarian cancer between 10% and 20%. Many people prefer not to know if they have a genetic mutation linked to cancer. Some physicians have stressed the importance of conveying information, not only about the benefits of testing, but about the limitations of the tests. Nuclear medicine tests can provide guidance to anxious patients by making it possible for physicians to monitor individuals with an increased risk of developing cancer This ability may influence the decision about whether to perform the genetic tests.

Nuclear medicine can also help determine the relative importance of genetic and environmental factors in diseases such as mental illness, hypertension, diabetes, and other common diseases. For example, with ^{99m}Tc-Sestamibi we can now assess a person's ability to detoxify carcinogenic or other disease-causing agents. With other radiotracers, we can detect the earliest regional biochemical signs of disease in susceptible individuals long before they develop symptoms. In some cases, detection of the first signs of disease will permit prevention of more severe consequences by treatment at a preclinical stage. Physicians will increasingly deal with asymptomatic subjects, and not just the clinically sick.

Molecular nuclear medicine will make diagnosis more specific. Instead of treating the patient on the basis of statistical responses, each patient will have his or her specific regional molecular diagnoses.

An excellent example of the interplay of genetics and nuclear medicine is the question of why so many types of cancer manifest an increased accumulation of fluorine-18 deoxyglucose (FDG). High FDG uptake is often the result of increased expression of type II hexokinase, associated with anaerobic metabolism of glucose. There may also be overexpression of GLUT-1, the glucose transporter in the cancer cell membrane (Schoenberger et al. 1996). On the other hand, in studies of mice with tumor xenografts, Aloj, Neumann, and colleagues from the National Institutes of Health (Aloj et al. 1996) found that there was a poor correlation between GLUT-1 mRNA expression and the amount deoxyglucose accumulation.

2.6 Dementia Susceptibility Genes

Three diseases illustrate the relationship between genotype and molecular phenotype: Alzheimer's, Lesch-Nyhan syndrome, and Rett syndrome. When genetic risk assessment (apolipoprotein E-4 allele APOE-4 mutation) was combined with FDG-positron emission tomography (FDG-PET) in relatives at risk for familial Alzheimer's disease, those family members with APOE-4 had the same abnormal patterns of reduced temporoparietal FDG accumulation that is characteristic of patients with Alzheimer's disease (Small et al. 1996). They had these patterns even though they were asymptomatic and several years younger than the average age of onset of Alzheimer's. This finding is important for selection of patients for early trials of putative Alzheimer's-prevention drugs and for providing persons at risk with a means of early diagnosis by periodic testing.

2.7 Inherited Metabolic Disorders

A drastic decrease in regional brain oxygen metabolism where regional cerebral blood flow was preserved was found in patients with a mitochondrial gene defect that limits the human brain to anaerobic glucose utilization. In patients with the same disease but with lactic acidosis, cerebral glucose utilization was normal.

Phenotypic characterization of genetic abnormalities is not limited to the brain. Huang and colleagues from the National Defense Medical Center in Taiwan measured the time–activity curves over the liver to distinguish two causes of hyperbilirubinemia: Dubin-Johnson syndrome, in which the genetic defect is in transport of bile into the collecting system, and Rotor syndrome, where the defect is in transport of bilirubin from plasma to hepatocytes (Huang et al. 1996).

2.8 Monitoring Gene Therapy

Herpes simplex virus (HSV), which has been genetically altered to express thymidine kinase, can enter cancer cells and increase the expression of the enzyme thymidine kinase.

Radiolabeled antiviral antibiotics are phosphorylated by thymidine kinase, and the product remains trapped in the cancer cells. Investigators at UCLA synthesized fluorine-18 acyclovir in order to assay the thymidine kinase induced by HSV as a measure of the effectiveness of entry of HSV into the cancer cells (Srinivasan et al. 1996).

Goldman and colleagues from Cliniques Universitaires de Bruxelles (Goldman et al. 1996) from Belgium showed that fluorine-18 FHPG, an analogue of gancyclovir, entered tumor cells and was a promising radiotracer for the evaluation of gene therapy involving viral thymidine kinase genes. Haberkorn (Haberkorn et al. 1996) from the German Cancer Research Center in Heidelberg found that the thymidine kinase-expressing hepatoma cells accumulated deoxyglucose, which the researchers attributed to a "stress" reaction of the tumor cells.

References

Aloj L, Caraco C, Jagoda, E, Moody T, Eckelman WC, Neuman RD (1996) Facilitative and sodium dependent glucose transporter MRNA expression in nude mice tumor xenografts: correlation with deoxyglucose uptake. J Nucl Med [Suppl] 37:254P

Bom HS, Kim YC, Song HC, Kim JY, Park KO (1996) Tc99m sestamibi uptake in small cell lung cancer: a predictor of response to chemotherapy. J Nucl Med [Suppl] 37:67P

Franssen EJF, Kool W, Bakker M, van der Graaf WTA, Hollema H, Groen HJM, Müller E, Smit EF, deVries EGE, Piers DA (1996) Determination of functional p-glycoprotein in patients with solid tumors by technetium-99m sestamibi scintigraphy with and without PSC 833. J Nucl Med [Suppl] 37:67P

Goldman S, Monclus M, Cool V, Damhaut P, Luxen A, Levivier M, Brotchi J, Hildebrand J, Velu T (1996) A novel PET tracer for evaluation of gene therapy. J Nucl Med [Suppl] 37:53P

Haberkorn U, Altmann A, Morr I, Germann C, van Kaick G (1996) Metabolic effects of gene therapy with HSV thymidine kinase in hepatoma cells. J Nucl Med [Suppl] 37:234P

Huang WS, Yung TH, Guan SI, Chen WL, Tung HS (1996) Kinetic cholescintigraphy in the evaluation of patients with Dubin-Johnson and Rotor syndromes. J Nucl Med [Suppl] 37:247P

Kapucu LÖ, Akyüz C, Vural G, Oguz A, Atasever T, Büyükpamukcu M, Ünlü M (1996) The value of MIBI scintigraphy in predicting the prognosis in pediatric patients with lymphoma. J Nucl Med [Suppl] 37:139P

Schoenberger JA, Rueschoff J, Marienhagen J, Meyringer R, Bock E, Hofstadter F, Eilles C (1996) Correlation of increased glucose transporter 1 gene expression and the degree of histological differentiation in thyroid neoplasms. An immunohistochemical study. J Nucl Med 37:248P

Sharma V, Wey SP, Bass L, Crankshaw C, Green MA, Welch MJ, Piwnica-Worms D (1996) Monocationic N_4O_2 schiff-base phenolate complexes of Ga(III): novel PET imaging agents of the human multidrug resistance (MDR1 P-glycoprotein. J Nucl Med [Suppl] 37:51P

Small GW, Saxena S, Mazziotta JC, Barrio JR, Cole GM, LaRue A, Mandelkern MA, Saunders AM, Haines JL, Pericak-Vance MA, Roses AD, Phelps ME (1996) Strategies using PET for early detection of Alzheimer disease. J Nucl Med 37:79P

Srinivasan A, Gambhir SS, Green AL, Cherry SR, Sharfstein S, Barrio JR, Satyamurthy N, Namavari M, Wu L, Berk AJ, Phelps ME (1996) A PET reporter gene (PRG)/PET reporter probe (PRP) technology for repeatedly imaging gene expression in living animals. J Nucl Med [Suppl] 37:107P

3 Integrated Imaging of the Heart: New Clinical and Experimental Approaches

E.A. Zerhouni

3.1 Introduction 23
3.2 Heart Function 24
3.3 Magnetic Resonance Imaging Methods 24
3.3.1 Imaging of Global and Local Mechanical Function 26
3.3.2 Imaging of Tissue Perfusion 28
3.3.3 Imaging of the Coronary Vessels 30
3.4 Image Analysis 31
3.5 New Directions 31
Reference 32

3.1 Introduction

A comprehensive understanding of the heart in health and disease requires detailed knowledge of multiple functional parameters such as integrity of the cardiac anatomy, adequacy of vascular supply during both rest and stress conditions, local and global mechanical function, and, ideally, the underlying metabolism. These parameters are now obtained through a multiplicity of separate methods, including echocardiography, radionuclide imaging, coronary angiography, and magnetic resonance imaging (MRI).

3.2 Heart Function

Harmonious contraction of myocardial fibers for efficient ejection of blood from the ventricular chambers is the ultimate functional goal of the heart. Accurate study of myocardial contraction is difficult because of the complex motion of the heart during its cycle. Imaging methods that rely either on projectional or on tomographic images cannot display the relative displacement of specific myocardial points during the entire contraction cycle.

3.3 Magnetic Resonance Imaging Methods

Given the lack of distinctive anatomic landmarks, it is impossible, for instance, to ensure that end-diastolic (ED) and end-systolic (ES) images of a short-axis plane represent the same myocardial locus, because through-plane motion or in-plane shear cannot be detected. Thus any measure of myocardial mechanical function, such as radial thickening, is inherently flawed. This problem has long been recognized, and methods based on the physical implantation of metallic markers into the myocardium have been extensively used in the elucidation of ventricular mechanics. However, these methods are invasive, disturb the physiology under study to an unknown extent, and are limited to small regions of the myocardium. With the advent of MRI, noninvasive, high-resolution, and spatially registered images of the entire heart throughout its contractile cycle can be obtained. Nevertheless, conventional MRI suffers from the same limitations as other tomographic imaging methods: the complex cardiac motion patterns and the paucity of natural landmarks prevent "true" analytical description of myocardial function. Methods able to unambiguously encode the intrinsic motion of material points of the myocardial wall would thus be of great value in the characterization of heart function as well as contributing to a comprehensive, integrated cardiac examination.

The intrinsic multiplanar capabilities of MRI allow easy evaluation of anatomically complex structures. The intrinsic sensitivity of MRI to motion effects can be used advantageously to study flow patterns in vascular structures, including the coronary arteries. The high inherent tissue contrast allows sensitive detection of both endogenous and exogenous contrast agents in the intact human subject. In addition to the standard depiction of myocardial contraction, novel methods using either tissue tagging or multidirectional phase encoding permit unique calculation of three-dimensional deformation within the myocardial tissue during its entire contractile cycle.

Despite its enormous theoretical advantages, MRI has not yet achieved wide clinical acceptance because current scanners cannot perform a comprehensive cardiac study within reasonable examination times. Recent research and development of high-performance hardware and pulse sequences, however, now permit the formulation of more effective study protocols. Faster imaging sequences requiring high-performance gradients have now been implemented successfully, and most manufacturers have improved the performance of imaging gradients to the point where routine, clinically effective, high-speed acquisitions can be obtained. This has been greatly enhanced by the use of innovative pulse sequences such as gated single-breathhold segmented K-space techniques and echo planar imaging (EPI). Although these advances have significantly improved the prospects for successful implementation of an integrated and time-efficient cardiac MRI examination, current scanners are still hampered by inadequate software design and slow computer systems, which are being redesigned to allow near real time interactive scanning. A consensus is now emerging indicating that dedicated high-speed scanners optimized for cardiovascular applications will be essential to the clinical success of cardiac MRI.

The success of an integrated protocol depends, in great part, on the utilization of pulse sequences that are flexible enough to allow cine imaging in multiple arbitrary orientations from a single prescription. In addition, these sequences should permit detailed quantitative motion studies using either myocardial tagging or multidirectional phase encoding. The same cine sequence modified in real time at the switch of user-defined control variables can also be used for assessment of perfusion using exogenous contrast agents and coronary imaging. These sequences should allow single breathold acquisitions and be compatible with multicoil array imaging for maximum signal-to-noise ratio (SNR). We have implemented this approach by developing a "master" cardiac imaging sequence based primarily on the segmented K-space approach initially proposed by Atkinson and Edelman. An optimization algorithm for arbitrary imaging angles was added to the sequence, and a user interface allows the operator to activate a series of imaging parameters (e.g., the number of phase-encoding steps and their groupings, the addition of a tagging pattern in the X , Y, or Z directions) without having to reload the sequence software. For perfusion study, a single command can be activated allowing acquisition of perfusion images following preparation by a train of radiofrequency (RF) pulses to reach steady state. This master sequence includes spectral fat saturation pulses to enable coronary imaging to be performed as well as slab saturation to eliminate signal from inflowing blood when necessary. The development of such integrated master cardiac sequences is a key to the successful implementation of a comprehensive clinical cardiac examination.

Our current examination strategy comprises three separate steps:

1. Examination of mechanical function at rest and during pharmacologic stress
2. Performance of an exogenous contrast-enhanced series using clinically approved agents
3. Study of the proximal coronary vessels focusing on the coronary territories suspected of being abnormal as determined by steps 1 and 2

All studies are performed with a specially developed cardiac multicoil array for maximum SNR. On-line echocardiography (ECG) and blood pressure monitoring as well as pulse oximetry are always used. An indwelling intravenous catheter with saline perfusion via a calibrated MRI-compatible infusion pump is inserted prior to the examination.

3.3.1 Imaging of Global and Local Mechanical Function

ECG-gated segmented K-space cine acquisitions during suspended respiration provide views of the heart during systole and diastole at temporal resolution of 30 to 60 ms. Complete three-dimensional coverage of the myocardium in long and short axes can be accomplished using 12–18 breatholds lasting 16–24 heartbeats with a standard gradient system and three to six breatholds with multishot EPI capability. All of the cine acquisitions are tagged using a DANTE-SPAMM scheme in a set of six parallel short-axis images and a set of six radially distributed long-axis images, with a matrix of 256 × 110 and five phase groupings with a repetition time (TR) of 6.5 ms, an echo time (TE) of 2.3 ms, and a flip angle of 15°–20°. Slab saturation is applied parallel to the image plane to decrease the signal of ventricular blood. Because tag persistence is dependent on the T_1 relaxation time of myocardium, these image series are always obtained prior to injection of any contrast medium which shortens T_1 of myocardium.

With segmented multishot EPI, the acquisition of images is fast enough to allow both rest and post-dobutamine stress imaging with stages of 5 min each. At the present time, for validation and research purposes, we select patients who have already undergone dobutamine echo stress testing and we do not exceed the dose of dobutamine prior to that at which wall motion abnormalities were observed with ECG. Others have shown the feasibility of such studies with standard scanners. However, due to the requirement for on-line monitoring of wall motion to detect ischemia, we believe that such studies require high-speed acquisition capabilities.

These cine-tagged images are viewed on-line for qualitative evaluation of myocardial contraction while the patient is being examined. The cine images are then analyzed off-line. After semiautomated segmentation of epicardial, endocardial contours and linear tags, a three-dimensional finite element analysis of the *true* material deformation of the myocardium can be performed and spatially coregistered with the image data to provide a context-sensitive anatomic and functional data set. Usual parameters of global function, such as global and regional ejection fraction, ventricular mass, and wall stress, can then all be easily extracted. Three-dimensional reconstruction at each time point permits dynamic display of all strain parameters independent of global cardiac motion in a variety of formats. Although myocardial tagging or three-dimensional phase encoding are the only existing methods that permit truly noninvasive quantification of both global and local functional parameters, the issue of whether such a detailed analysis of mechanical function is of greater clinical value than simpler semiquantitative planar methods is still being investigated. Nonetheless, with improved scanners and image analysis methods, it is likely that the ability to quantify local function more accurately may find use in the development of more objective and operator-independent methods of measuring function during rest and graded levels of stress. The development of such an inherently *quantitative stress test* could provide a more sensitive and objective method of detecting local ischemia than two-dimensional stress echocardiography, which suffers from the lack of spatial coregistration of myocardial regions at rest and stress due to the change in the global shape of the heart during inotropic stress. The ability to quantitatively measure strain of the same *material* points rather than the same *image* points with MRI at rest and stress is an unprecedented capability. We have preliminarily shown that ischemic wall dysfunction can be demonstrated at lower levels of pharmacologic stress than with echocardiography. In addition, recovery of function in the regions surrounding infarcts can also be seen in the three-dimensional MRI strain data at low-dose dobutamine stress, thus identifying stunned but viable myocardium.

Combined with concurrent measures of local perfusion at spatial resolution greater than that achieved with radionuclide methods, such studies form a potentially powerful and cost-effective approach to the characterization of ischemia.

Newer mechanisms are still being sought for reaching even deeper levels of understanding of heart function. For example, the relationship of fiber arrangement and rearrangement to the function of the myocardium is poorly understood. It has been suggested more recently that diffusion sensitive sequences could provide a direct measure of myocyte orientation in vivo. With greater temporal resolution routinely achievable, studies of electrical activation have now started in the experimental animal and may lead to a better understanding of myocardial excitation and remodeling mechanisms following injury. Clearly, functional cardiac MRI is an actively evolving field with enormous clinical implications and significant prospects for elucidating basic questions in cardiac disease.

3.3.2 Imaging of Tissue Perfusion

At the present time, the most reliable results in clinical scanning are obtained with gadolinium-based contrast agents using the T_1-shortening effect of the agent with T_1-sensitive pulse sequences. All of the clinically approved MRI contrast media distribute freely within the vascular and extravascular extracellular spaces with a T_1-shortening effect linearly proportional to the concentration of gadolinium. Accurate in vivo quantitation of the local concentration of the agent via MR images is, however, unlikely to be achieved because of exchange rate considerations between vascular and interstitial space and the complex influence of imaging parameters on signal intensity. Nonetheless, it appears that a semiquantitative approach may provide sufficient information for the sensitive detection of functionally ischemic regions prior to infarction and for the assessment of reperfusion after infarction.

Our protocol for imaging of myocardial perfusion uses an image sequence that maximizes T_1 contrast while preserving dynamic range and linearity of response to the concentration of gadolinium in both the ventricular cavity and the myocardium. To achieve these goals, the magnetization is first driven to saturation by a series of preparatory nonselective RF pulses followed by a segmented K-space spoiled gradient echo imaging sequence with a flip angle of 45° and a matrix of 256×96 phase-encoded steps acquired in the same short-axis planes as the functional images described above. By optimizing the RF saturation pulses, it is possible to obtain two to three separate slices (depending on heart rate) in three successive heartbeats using 32 phase-encoded steps per heartbeat per image plane on a standard gradient system. Using multishot EPI, we can acquire six image planes in two to four heartbeats. Images are acquired as rapidly as possible following IV injection of the contrast medium. For the detection of suspected ischemia, contrast injection is performed immediately after the last stage of cine imaging during pharmacologic stress.

In patients with coronary flow restriction, transiently delayed opacification in the endocardial layer of the hypoperfused region is observed. Our preliminary observations indicate that these zones of delayed enhancement correlate highly with the presence of flow-restricting lesions and that the amount of delay may predict the severity of the lesion. Encouragingly, these transient phenomena in the endocardium appear to last for periods of up to 30–60 s and are thus largely within the temporal resolution of current perfusion sequences.

In patients with recent established infarction, contrast agents provide insight into the status of tissue integrity in the ischemic region as well as an index of the effectiveness of reperfusion measures. Central hypo-enhancing zones that persist for several minutes in the infarct region have been shown to correlate with the "no reflow" phenomenon in experimental canine models. Late hyper-enhancing regions, on the other hand, have been shown to correlate with the triphenyltetrazolium (TTC)-negative infarcted, nonviable myocardium both in humans with reperfused infarcts and in a canine model of infarction–reperfusion.

3.3.3 Imaging of the Coronary Vessels

An important aspect of the cardiac MRI study is the ability to accurately coregister functional data in three dimensions. For instance, the perfusion images can be superimposed on the mechanical function images into a single function/perfusion map, since the same imaging planes are obtained for both. This capability is likely to enhance interpretation and improve accuracy of diagnosis. In addition, the finding of matched abnormal regions on tagged cine studies and perfusion studies in the same patient increases the likelihood of finding significant coronary flow restriction, which may ultimately lead to better patient selection for invasive coronary angiography. During the MRI examination, it also allows the MR coronary angiogram to be tailored to the coronary vessel most likely to be responsible for the detected abnormality. This "guided" approach to coronary imaging by MR is, we believe, warranted because of the still unreliable nature of MR coronary angiography. In our experience, it has been difficult to obtain satisfactory and complete depiction of all coronary vessels in one setting. The image resolution is still suboptimal for reliable, "stand-alone" diagnosis of coronary lesions. The combination of functional and anatomic information in specific territories is, however, effective since the functional significance of any lesion can be assessed. Using surface coils, the proximal 5–8 cm of the right coronary artery can be visualized in over 90% of patients, whereas success rates are more variable for the left main, left anterior descending, and circumflex arteries. Furthermore, lower success rates are experienced in female patients owing to the smaller diameter of coronary arteries in women.

Scanning at suspended respiration during diastole using fat suppression with the scan plane oriented tangentially to the myocardial surface and a thick slice of 10–20 mm, a 256×192 matrix, a flip angle of 15°–20°, and a field of view of 24–28 cm with 16 phase-encoded groupings is the method we prefer to use. This method allows direct visualization of the coronary vessel without the need for vessel segmentation from thin, contiguous slices. This "tangential" scanning approach permits visualization of longer coronary segments in the region of suspected abnormality. Additional infusion of contrast material is sometimes required for better visualization of more distal segments of the targeted vessel. At the present time, MR coronary angiography remains the most challenging portion of the integrated cardiac MR examination, and further progress is required for widespread clinical applicability.

3.4 Image Analysis

The adequacy of image analysis protocols for cardiac MRI remains a challenge. Off-line analysis still requires both time and operator interventions that limit the rapid availability of the information. Because of the complex and extensive amount of information generated during cardiac MRI, more efficient forms of data display and visualization using state-of-the-art graphics computers are actively being developed. A promising approach, uniquely suited to MRI data, is the interactive display of both dynamic anatomic data and functional data derived from mesurements of strain, perfusion, and coronary flow.

3.5 New Directions

In the research laboratory, MRI tagging is fast replacing or complementing all other existing methods for such purposes. Whether MRI tagging will provide useful incremental knowledge for clinical applications is actively being tested. It is likely, however, that even if MRI tagging provides significant functional information, its future in the clinical arena will be determined by the successful implementation of other MRI methods geared toward direct assessment of the coronary tree and myocardial perfusion, combined in a single, comprehensive, and cost-effective MRI examination.

It is clear that a period of increased experimentation will lead to rapid changes in the approach to the single-step comprehensive cardiac MRI examination. The exploitation of novel contrast mechanisms such as blood oxygen-dependent contrast (BOLD) may lead to more direct measurements of oxygenation at the tissue level. Perfusion measurements by spin-tagging techniques also appear promising and may find a place in the examination protocol of the future. New contrast agents using T_2 contrast mechanisms or purely intravascular agents may play a prominent role. Cardiac spectroscopy, which is still a daunting technical challenge, may provide MRI with a metabolic dimension of great clinical value. The operator-independent nature of MRI examinations and the ability to quantify accurately multiple parameters of flow, perfusion, function, and anatomy at spatial and temporal resolutions unmatched by any other existing method is very likely to result in a cost-effective, integrated cardiac imaging process of high diagnostic value.

Reference

Atkinson DJ, Edelman RR (1991) Cineradiography of the heart in a single breath hold with a segmented turboFLASH sequence. Radiol 178: 357–360

4 Principles and Clinical Applications of Functional Brain Imaging

T.J. Brady, A.G. Sorensen, and B.R. Rosen

4.1	Introduction	33
4.2	Principles	34
4.3	Brain Tumors	36
4.4	Presurgical Planning	38
4.5	Acute Stroke	39
References		41

4.1 Introduction

Functional imaging techniques are well established in both the nuclear medicine and positron emission tomography (PET) communities. While the informational content of these images can be high, limited availability (PET) and low spatial resolution limit their clinical applications. Magnetic resonance (MR) imaging can provide both functional and anatomic information, in part because of the large number of intrinsic parameters that affect MR image contrast. These include proton density (r), relaxation times T_1 and T_2, chemical shift, flow, magnetic susceptibility, diffusion, and others. While not functions of tissue per se, each of these parameters can be exploited to assess various functional activities of the body. This is best illustrated by MR angiography, where both time-of-flight and phase contrast methods depend on velocity (cardiovascular function) to provide contrast and thereby to represent vascular anatomy. Similarly, it can be argued that a positive gadolinium-diethylenetriaminopentoacetic acid (Gd-DTPA) central nervous system (CNS) examination reflects abnormal function (or integrity) of the blood–brain barrier.

Synergistic developments in MR imaging instrumentation, image acquisition strategies, and MR contrast agents have increased the availability of techniques for and approaches to functional MR imaging. Refinements in fast MR imaging techniques, e.g., fast gradient echo and echo-planar imaging (EPI), have been especially important in this regard, given their success in reducing motion artifacts and their ability to increase the number of studies performed per patient examination (Cohen and Weisskoff 1990). Of particular interest, real-time imaging methods essentially freeze action and thereby allow physiological motion, such as cardiac systole, to be evaluated. Most recently, these techniques have been used to obtain accurate assessments of tissue and organ hemodymanics, e.g., blood volume and flow, by using both exogenous and endogenous contrast strategies. While the term functional MR (fMR) imaging commonly refers to mapping of signal changes with some type of task paradigm, we use the term in its fuller sense, including imaging water mobility (via diffusion-weighted imaging, DWI), permeability of the blood–brain barrier, and maps of relative cerebral blood flow (rCBF) and volume (rCBV).

4.2 Principles

The assessment of cerebral hemodynamics has been demonstrated using several MR approaches, including the following (Rosen et al. 1990; Kwong et al. 1992; Detre et al. 1992; Edelman et al. 1994):

1. Serial imaging during the first-pass transit of MR contrast agents
2. Change in oxy-deoxyhemoglobin concentration during neuronal activation
3. Novel MR pulse sequences to directly measure perfusion

The first two rely on magnetic susceptibility effects and require T_2-weighted pulse sequences.

Susceptibility-based perfusion imaging uses EPI to visualize the rapid passage of a bolus of contrast agent. All MR agents have both T_1

and T_2 relaxation effects. While routine clinical use of Gd-based agents depends on their T_1-relaxing effects (causing increased signal intensity on T_1-weighted images), perfusion-sensitive imaging uses the T_2 effects, which dominate at higher concentrations. The passage of contrast agent causes a transient signal loss (T_2 effects) proportional to the blood volume in each voxel. Thus this approach is sensitive to tissue microvasculature. rCBV maps can be generated by applying susceptibility physics and standard tracer kinetic principles. As with other intravascular tracer techniques, more than rCBV can be computed from the signal intensity–time curve. Contrast agent arrival time, time to peak change, and transit time can be calculated on a voxel-by-voxel basis. Over the past 5 years, perfusion imaging has demonstrated utility in various CNS applications, including tumor detection and/characterization and assessment of acute stroke[1].

Equally exciting is the non-contrast-agent approach that demonstrates real-time changes in MR signal in response to various neurological task activations. This blood oxygen level-dependent contrast (BOLD) approach exploits the decrease in arteriovenous (AV) oxygenation difference that accompanies regional increase in blood flow. This method utilizes changes in the levels of deoxyhemoglobin (the body's own contrast agent) to provide endogenous MR contrast (Ogawa et al. 1990). This increase in blood flow simultaneous with a reduction in deoxyhemoglobin concentration produces regional signal enhancement in the area of activation. In addition to cognitive neuroscience research, this approach is being used in presurgical planning in patients with CNS lesions.

Diffusion is the random, thermal motion of molecules (also known as brownian motion). The quantity D is a measure of diffusion and is typically expressed in units of an area per unit time. The D for pure water at room temperature is about 2×10^{-5} cm^2/s. MR imaging can be made sensitive to intravoxel dephasing, and therefore diffusion, by the addition of gradient pulses to a standard spin-echo sequence. MR imaging has provided one of the first noninvasive in vivo methods for

---

[1] Strictly speaking, "perfusion imaging" is a misnomer, since contrast injection studies typically do not measure flow directly, but rather other hemodynamic parameters. Efforts are in progress to obtain true cerebral blood flow measurements using MR contrast agents.

measuring a diffusion coefficient, which is known as the apparent diffusion coefficient (ADC). While diffusion and brain function may not appear to be closely linked, studies have demonstrated that the apparent diffusion coefficient changes in disease states.

4.3 Brain Tumors

Diagnosis of CNS tumors has been greatly aided by the anatomic and contrast resolution of MR imaging. Characterization is more difficult, particularly because the T_1 and T_2 relaxation times of neoplastic tissue are not specific indicators of malignancy. PET studies of tumor metabolism using fluorodeoxyglucose-18 (^{18}FDG) have shown the degree of glucose utilization to be a predictor of tumor prognosis. However, metabolism may not be the only or the best method to evaluate cancer. In order to continue growth, a tumor must induce new capillary vessels once it reaches a few millimeters in size; this is termed angiogenesis. Additional investigation has demonstrated that tumor angiogenesis is an independent predictor of relapse-free survival in primary breast carcinoma. Susceptibility-based MR perfusion imaging, with its sensitivity to the capillary bed, is therefore well suited for evaluating tumor angiogenesis in vivo.

The susceptibility contrast agent MR approach has several features that are important for use in patients with neoplastic disease (Aronen et al. 1994). First, rCBV maps appear closely correlated to tumor grade, with low-grade lesions showing low rCBV (low MR signal) and more aggressive lesions demonstrating elevations of rCBV (increased MR signal). This is in accordance with the results of previous angiographic and PET studies, which found increased rCBV when studying high-grade primary brain tumors. Second, in some lesions, significant regional heterogeneity of rCBV was apparent. High-grade lesions in particular often showed considerable variation across the tumor. In lower grade lesions, focal high rCBV regions were often not apparent on conventional pre- or postcontrast T_1- and T_2-weighted images. These cases demonstrate that functional rCBV imaging may have an important role in guiding tumor biopsy. Despite the excellent correlation, interesting differences were apparent across the brain between these two modalities. Notably, MR rCBV images show relative insensitivity to large

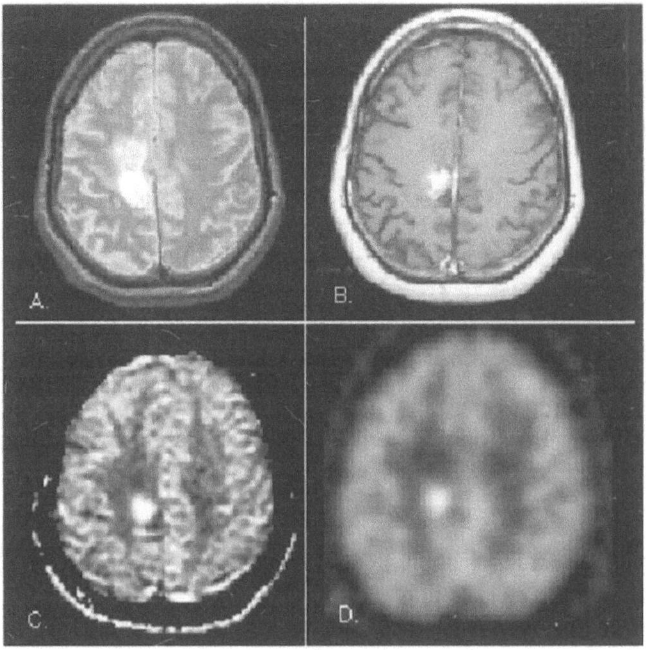

Fig. 1A–D. A 47-year-old male with known low-grade oligodendroglioma status after treatment, now with new enhancement on routine MRI. **A** T_2-weighted image. **B** Post-Gd T_1-weighted image. **C** Relative cerebral blood volume (rCBV) map showing high rCBV focus. **D** Fluorodeoxyglucose-18 (^{18}FDG) positron emission tomography (PET) showing increased metabolic uptake. Biopsy showed anaplastic astrocytoma

vessels such as the middle cerebral artery and central veins. This reflects the increased sensitivity of the MR approach to the microvasculature.

A third finding was quite surprising. In the majority of our tumor cases studied with both MR rCBV and PET ^{18}FDG, there was good to excellent congruence between these maps (Fig. 1). With both imaging modalities, a focal "hot spot" is seen, showing congruence between the region of increased glucose uptake and the region of increased microvascular volume. The origin of this relationship is currently being explored. Based on our results and previous studies, we postulate a link between tumor metabolism, tumor angiogenesis and microvascular den-

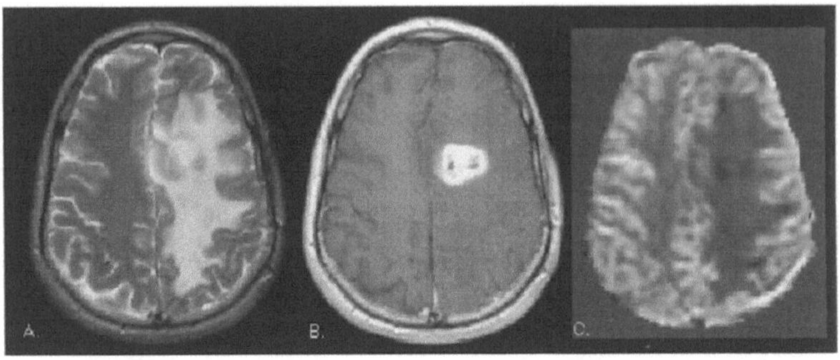

Fig. 2A–C. A 42-year-old male with known grade 3/4 glioma status after radiation therapy, now with enlarging mass and new enhancement. **A** T_2-weighted image. **B** Post-Gd T_1-weighted image. **C** Relative cerebral blood volume (rCBV) map showing no increased rCBV. Biopsy showed radiation necrosis only, with no evidence of malignancy

sity, and tumor grade. Finally, MR rCBV maps were able to differentiate recurrent tumor from radiation necrosis (Fig. 2).

Based on our experience (500 patients), we conclude that MR imaging blood volume mapping will be useful in the classification process of gliomas, in selecting optimal biopsy sites, in separating radiation necrosis from areas of tumor regrowth, and in the planning of radiotherapy. Moreover, the finding that glucose uptake and capillary blood volume are closely correlated from tumor to tumor and between regions within the same tumor may assist our understanding of the mechanisms regulating tumor angiogenesis and its link with tumor energy demands. The use of high-speed echo-planar imaging techniques, which provide the necessary temporal resolution, are essential for these studies of the entire brain.

4.4 Presurgical Planning

Forty patients with lesions near language and/or sensorimotor areas were studied by both fMR imaging and intraoperative direct cortical stimulation. Lesions included primary brain tumors, cavernous malformations, cortical heterotopia, and indeterminate pathologic diagnosis.

All patients underwent sensorimotor activation studies and/or language activation. Functional MR imaging was performed using the susceptibility-based BOLD contrast technique at 1.5T. Multislice T_2*-weighted echo-planar images were obtained using an asymmetric spin-echo pulse sequence during alternating control and activation tasks. Sensorimotor tasks consisted of self-paced repetitive movements of the fingers, toes, or tongue. Language tasks included verb generation, passive reading, and object recognition. All language tasks were performed without overt speaking in order to minimize motion artifacts. Sites of functional activation were identified by statistical analysis of the signal time course on a voxel-by-voxel basis using the Kolmogorov-Smirnov test. The functional maps were coregistered, fused, and volume rendered with high-resolution structural MR images. Multiple measurements of relative rCBF were also made by PET using inhaled $^{15}O\text{-}CO_2$ during rest and performance of the same tasks. Intraoperative primary sensorimotor and language mapping were performed by direct cortical stimulation using a hand-held bipolar stimulator (all cases) or depth electrode (one case) and were documented by photographs.

Volume rendering of the registered functional and structural images effectively demonstrated the three-dimensional relationships among cortical surface topography, tumor, and sites of functional activation. Each task within each subject activated localized regions in the contralateral precentral and postcentral gyri on both PET and fMR imaging, concordant with previous reports in normal subjects. Overall, there was good concordance between functional localizations by fMR imaging, PET, and direct cortical stimulation. fMR imaging may eliminate the need for the Wada test for language localization and reduce or eliminate the time at surgery for direct cortical stimulation studies.

4.5 Acute Stroke

While conventional computed tomography (CT) and MR imaging are excellent modalities for detecting and characterizing CNS disease in general, they fail to reliably detect acute ischemia or infarction at its earliest stages. DWI has been shown to be sensitive to early ischemic changes in brain (Moseley et al. 1990b; Warach et al. 1992), but does not directly detect ischemia itself as susceptibility-based perfusion imaging does (Kucharczyk et al. 1993). DWI has also demonstrated abnormalities in conditions other than ischemia animal models and therefore may lack specificity in humans. Using EPI, a variety of groups have developed and tested DWI (Le Bihan et al. 1992; Chien et al. 1992) and perfusion-sensitive techniques (Villringer et al. 1988; Moseley et al. 1990a) in humans and animals with cerebral ischemia. We have developed a protocol that acquires T_1, T_2, diffusion, MR angiography, blood–brain barrier (BBB) integrity, and perfusion-sensitive imaging in 30 min (14 min of data acquisition) in the setting of acute stroke (Sorensen et al. 1996).

Eleven patients were imaged with an average time from onset of symptoms of 6 h (range, 2–10 h). The diagnosis of acute cerebral infarction was confirmed in nine of 11 patients by follow-up clinical and imaging studies; the symptoms of the other two patients resolved quickly and had negative follow-up examinations. In the nine patients with infarct, eight had normal CT and seven had normal MR examinations in the acute phase; one patient showed subtle loss of the gray–white junction on CT at 3 h post-ictus, and two patients had subtle gyral hyperintensity (at 3 and 7 h post-ictus) on MR imaging. Five patients had intravascular enhancement on post-Gd T_1-weighted imaging. In nine patients with infarct, the DWI and perfusion-sensitive MR examinations demonstrated clear abnormalities in the acute phase that were confirmed by follow-up conventional imaging. Eight of the nine patients with confirmed acute infarct demonstrated markedly reduced flow signal in the appropriate middle cerebral artery (MCA). However, the degree of flow signal reduction did not correlate with the size or location of the infarct.

DWI and perfusion-sensitive MR imaging were consistent with three general categories:

1. Abnormalities of similar size and location on acute and follow-up studies
2. Initial DWI smaller than perfusion-sensitive MR abnormality, with a final infarct size larger than either
3. Initial DWI smaller than perfusion-sensitive MR abnormality, with a final infarct size between that of the DWI and perfusion-sensitive MR abnormalities

Pathophysiologically, the first group may indicate there may not be additional tissue at risk for the current vascular occlusion: collateral flow will limit the infarct size to the lesion already seen on the DWI. The initial mismatch in the second pattern may represent a small infarct surrounded by additional tissue at risk; without effective treatment, the infarct grows over time to involve even more tissue, perhaps due to excitotoxic, inflammatory, or free radical mechanisms in addition to secondary vascular mechanisms. The third pattern also shows a mismatch, but eventual infarct size smaller than the initial perfusion-sensitive MR abnormality. This may represent successful limitation of the infarct size due to recruitment of collateral flow. Both of the patients who exhibited this pattern in our early data were kept hypertensive after their cerebral infarct for the purpose of maintaining collateral flow. The presence of perfusion-sensitive MR abnormalities without DWI abnormalities supports the work in animal models indicating that DWI alone may not be sufficient for determining the presence or extent of cerebral ischemia, since there appears to be a threshold effect before tissue becomes abnormal on DWI despite low rCBV. The combination of DWI with perfusion-sensitive MR imaging may help identify brain tissue still at risk for infarction in acute stroke patients and will therefore be more powerful than either of these two techniques used individually. With these tools to investigate the natural history of stroke, understanding of the pathophysiology of cerebral ischemia in humans should increase. This in turn should enable improved rational therapy design.

References

Aronen H, Gazit I, Louis D, Buchbinder B, Pardo F, Weisskoff R, Harsh G, Cosgrove G, Halpern E, Hochberg F, Rosen B (1994) Cerebral blood volume maps of gliomas: comparison with tumor grade and histologic findings. Radiology 191:41–51

Chien D, Kwong KK, Gress DR, Buononno FS, Buxton RB, Rosen BR (1992) MR diffusion imaging of cerebral infarction in humans. Am J Neuroradiol 13:1097–1102

Cohen M, Weisskoff R (1991) Ultra-fast imaging. Magn Reson Imag 9:1–37

Detre J, Leigh J, Williams D, Koretsky A (1992) Perfusion imaging. Magn Reson Med 23:37–45

Edelman R, Siewert B, Darby D, Thangaraj V, Nobre A, Mesulam M, Warach S (1994) Qualitative mapping of cerebral blood flow and functional localization with echo planar MR imaging and signal targeting with alternating radiofrequency (EPISTAR). Radiology 192:1–8

Kucharczyk J, Vexler ZS, Roberts TP et al (1993) Echo-planar perfusion-sensitive MR imaging of acute cerebral ischemia. Radiology 188:711–717

Kwong KK, Belliveau JW, Chesler DA, Goldberg IE, Weisskoff RM, Poncelet BP, Kennedy DN, Hoppel BE, Cohen MS, Turner R, Cheng H-M, Brady TJ, Rosen BR (1992) Dynamic magnetic resonance imaging of human brain activity during primary sensory stimulation. Proc Natl Acad Sci USA 89:5675–5679

Le Bihan D, Turner R, Douek P, Patronas N (1992) Diffusion MR imaging: clinical applications. AJR Am J Roentgenol 159:591–599

Moseley ME, Mintorovitch J, Cohen Y et al (1990a) Early detection of ischemic injury: comparison of spectroscopy, diffusion-, T2-, and magnetic susceptibility-weighted MRI in cats. Acta Neurochir [Suppl] (Vienna) 51:207–209

Moseley M, Cohen Y, Mintorovitch J et al (1990b) Early detection of regional cerebral ischemia in cats: comparison of diffusion and T2 weighted MRI and spectroscopy. Magn Reson Med 14:330–346

Ogawa S, Lee TM, Kay AR, Tank DW (1990) Brain magnetic resonance imaging with contrast dependent on blood oxygenation. Proc Natl Acad Sci USA 87:9868–9872

Rosen B, Belliveau J, Vevea J, Brady T (1990) Perfusion imaging with NMR contrast agents. Magn Reson Med 14:249–266

Sorensen AG, Buonano FS, Gonzalez RG, Schwamm L, Lev MH, Huang-Hellinger FR, Reese TG, Weisskoff RM, Copen W, Look R, Finkelstein S, Davis TL, Moreira J, Rosen BR, Koroshetz WJ (1996) Evaluation of hyperacute stroke with combined multislice diffusion-weighted and hemodynamics echo-planar MR imaging. Radiology 199:391–401

Villringer A, Rosen B R, Belliveau J W, Ackerman JL, Lauffer RB, Buxton RB, Chao YS, Wedeen VJ, Brady TJ (1988) Dynamic imaging with lanthanide chelates in normal brain: contrast due to magnetic susceptibility effects. Magn Reson Med 6:164–174

Warach S, Chien D, Li W, Ronthal M, Edelman RR (1992) Fast magnetic resonance diffusion-weighted imaging of acute human stroke. Neurology 42:1717–1723

5 Molecular Probing and Imaging by Magnetic Resonance Spectroscopy: Possibilities for Imaging Gene Therapy

B. Ross, S. Bluml, E.R. Danielsen, and K. Kanamori

5.1	Introduction	43
5.2	Basics Aspects	44
5.2.1	Identification of Metabolites	44
5.2.2	Relative Concentration	46
5.2.3	Quantitation by Internal or External Reference	48
5.3	Improving Identification of Metabolites	48
5.3.1	Spectral Editing	48
5.3.2	Decoupling	48
5.3.3	Quantitation Verifies Assignments	50
5.3.4	Multinuclear Magnetic Resonance Spectroscopy	51
5.4	Specificity	53
5.5	Imaging of Receptor Expression	53
5.6	Enzyme Activity In Vivo	56
5.7	Conclusion	62
References		63

5.1 Introduction

The aim of this symposium was expressed by Prof. M. Schwaiger and Prof. W. Semmler as follows: "... Gene therapy will demand more sensitive and more specific imaging modalites." This and the challenging title of the symposium prompted us to reexamine the experience of in vivo magnetic resonance spectroscopy (MRS), particularly in humans. What have MRS and metabolite-specific imaging to offer in this arena? With one exception – the application of ^{31}P MRS to a mouse made transgenic for creatine kinase (Koretsky et al. 1989)[1] – we know of no direct attempt to monitor gene therapy with in vivo MRS. Nevertheless, a copious literature supports the view that MRS can successfully identify and assay the major products of gene expression, enzyme activities and metabolite concentrations, in living tissues (Ross and Michaelis 1994).

This paper is divided into three parts. The first part introduces the methods of in vivo MRS used to identify and quantify an important but limited range of metabolites. By focusing on human brain disease, the successful use of MRS for diagnosis and therapeutic monitoring of patients is discussed. In the second and third parts on receptor expression and in vivo enzyme rate(s), we explore the capabilities of in vivo MRS in addressing the key questions raised by the desire to image gene therapy.

We conclude that, despite limited spatial and temporal resolution, and a sensitivity limit in the range of 10^{-4} *M*, in vivo MRS has many of the characteristics of an ideal tool for imaging successful gene therapy.

5.2 Basic Aspects

5.2.1 Identification of Metabolites

MRS is highly specific for metabolite identification, as illustrated in Fig. 1. When the instrument is "tuned" to the resonanance frequency of hydrogen nuclei (also called protons or ^{1}H), the proton MR spectrum specifies methyl and methylene protons of ethanol by their chemical shift (1.2 and 3.75 ppm, respectively). Fine chemical structure is indicated by coupling of neighboring protons, which results in characteristic multiple peaks of differing intensities, at each location.

[1] M.C. Bianchi (personal communication), using in vivo ^{1}H MRS, recently identified a child lacking creatine and phosphocreatine (PCr) from the brain in whom MR images were normal. This may be a "human knockout" of creatine phosphokinase (CPK).

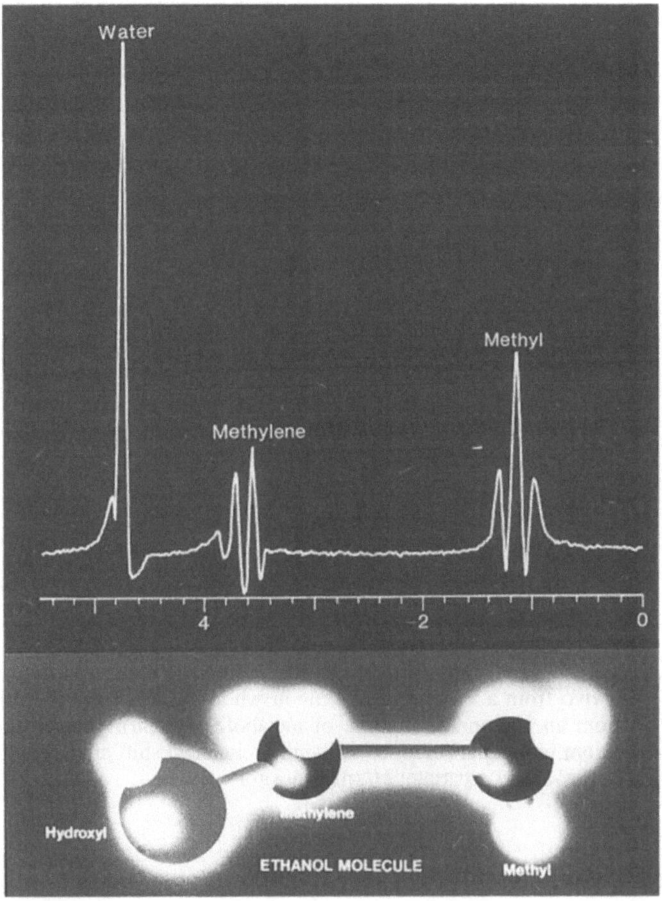

Fig. 1. Proton MR spectrum is a reflection of molecular structure. A simple molecule such as ethanol gives a complex MR spectrum (*top*), which allows the number of protons and the bonding structure (*bottom*) to be accurately deduced. The example shown is of a sample containing 20 m*M* ethanol in 1 l pure water. The spectrum shows the peaks of the methyl (-CH_3) and methylene (-CH_2) protons. The hydroxyl (-OH) proton resonates in the water peak and is not distinguished on this spectrum

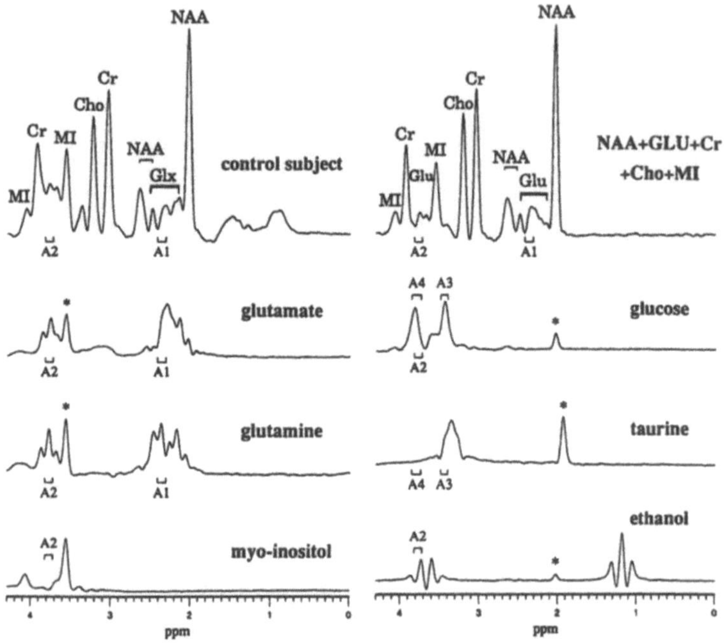

Fig. 2. Proton MR spectrum of in vivo human brain. The upper spectrum was obtained in vivo from a volume of 15 cm^3 in white matter of parietal cortex (*left*) and from an appropriate mixture of metabolites (*right*). The remaining spectra are from individual solutions to identify chemical shift and "coupling" patterns. *NAA*, *N*-acetyl aspartate; *MI*, myoinositol

5.2.2 Relative Concentration

The number of protons (3:2 in the case of ethanol) results in proportionate signal intensity, from which concentration can be determined if the concentration corresponding to any one signal in the spectrum is known. A quick method that is commonly used in in vivo MRS is to express peak areas as ratios to an arbitrarily selected one. In the in vivo human brain spectrum shown in Fig. 2, for example, peak identities and their relative concentrations can be very readily reconstructed by adding appropriate concentrations of pure solutions together in a flask (compare spectrum E, from a model solution, with the true in vivo spectrum

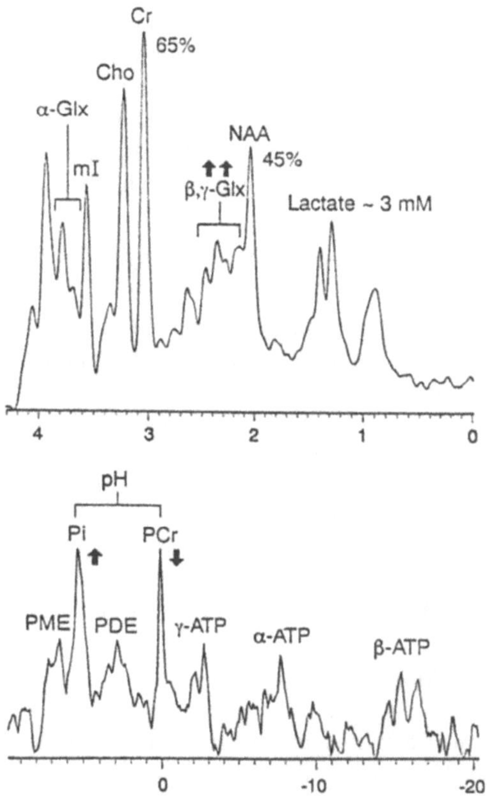

Fig. 3. In vivo human brain spectra at seven days after near-drowning (hypoxia) injury. *Above*, ^{1}H MRS. *Below*, ^{31}P MRS. *mI*, myoinositol. *Cr*, creatine; *PCr*, phosphocreatine; *PME*, phosphoethanolamine plus phosphoryl choline; *PDE*, glycerophosphoryl chline plus glycerophosphorylethanolamine *NAA*, *N*-acetyl aspartate; *Cho*, choline, *Pi*, inorganic phosphate. (See text for metabolite assignment and details)

in A). Furthermore, the subtly different coupling patterns of the crucial neurochemicals glutamine and glutamate permit their distinction with reasonable certainty in vivo (Fig. 2). In vivo brain ^{1}H MRS and constituents are shown as individual and combined "phantom" spectra.

Even quite small changes from normal can be monitored by in vivo ^1H MRS. This works well in sequential clinical studies. Making the working assumption that the intensity of Cr remains constant, Fig. 3 shows the progressive loss of *N*-acetyl aspartate (NAA) over time after neuronal injury as well as the excess of lactate, lipid, and glutamine that is expected as a late neurochemical response to hypoxia.

5.2.3 Quantitation by Internal or External Reference

True metabolite concentrations are not difficult to obtain from in vivo MRS examinations, but require a reference concentration (internal, such as water, or external in a vial mounted near the patient) or some calibration experiment to establish the stability and sensitivity of the MR scanner. Fortunately, MR scanners are now so reliable that comparable quantitative MRS data can be obtained anywhere in the world (Ernst et al. 1993; Webb et al. 1994).

5.3 Improving Identification of Metabolites

5.3.1 Spectral Editing

Serious errors in peak assignment are almost unavoidable in the crowded in vivo ^1H spectra obtained in routine clinical MRS. Changing the operating condition or the pulse sequence used to acquire the spectrum is often the simplest way to obtain confirmation of the assignment; examples are short versus long echo time and inversion of signal from coupled spins to identify lactate.

5.3.2 Decoupling

A particular example which provides valuable information about composite peaks in the ^1H and ^{31}P MR spectrum (Fig. 3, lower panel) is proton decoupling, increasingly applied in ^{31}P MRS of cancer and in the human brain (Bluml and Ross 1997).

The peak labeled "Cho" in ^1H MRS is a composite, containing, it was believed, phosphoryl choline (PC) and glycerophosphoryl choline

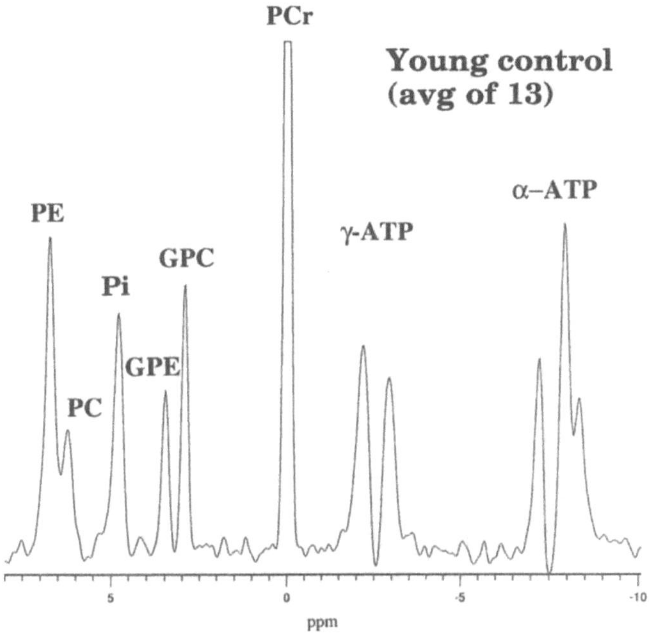

Fig. 4. Proton-decoupled ^{31}P MRS of human brain. A noninvasive, quantitative ^{31}P MRS examination performed in a routine clinical MR scanner. In this patient with chronic closed head injury, ^{1}H MRS showed persistent elevation of choline. Resolving complex monoester and diester peaks of the ^{31}P MR spectrum explains this by demonstrating an increase in brain glycerophosphoryl choline (*GPC*). *PCr*, phosphocreatine; *GPE*, glycerophosphorylethanolamine; *PE*, phosphoethanolamine; *PC*, phosphoryl choline; *Pi*, inorganic phosphate. (See text for details)

(GPC). The peaks labeled PME and PDE in ^{31}P MRS are also composite, PME containing PC and phosphoethanolamine (PE) and PDE containing GPC and glycerophosphorylethanolamine (GPE). By means of proton decoupling (Fig. 4) PME and PDE can be further resolved (note that elevated Cho in this case is probably due to excess GPC).

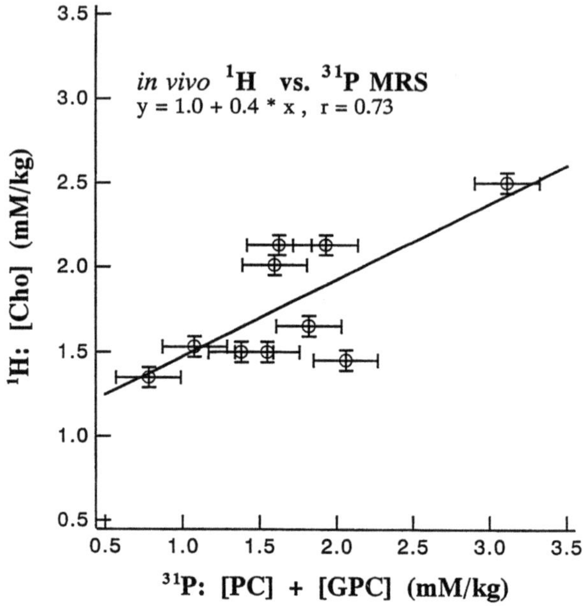

Fig. 5. Composition of cerebral "cholines." Comparing quantitative ^{1}H MRS with ^{1}H-decoupled ^{31}P MRS indicates from a positive intercept that additional metabolites probably contribute to the ^{1}H choline peak area

5.3.3 Quantitation Verifies Assignments

Accurate quantitation of both ^{1}H and proton-decoupled ^{31}P MRS adds another level of certainty to these assignments (Fig. 5) and at the same time reveals a new assignment. Closer examination of Fig. 5 shows that, despite a good linear correlation, the intercept is not zero. The assumption that all Cho of human brain is PC and GPC is incorrect; work in progress suggests that the strongly coupled resonances of GPE and PE quantitatively contribute most of the missing signal (Bluml et al. 1997).

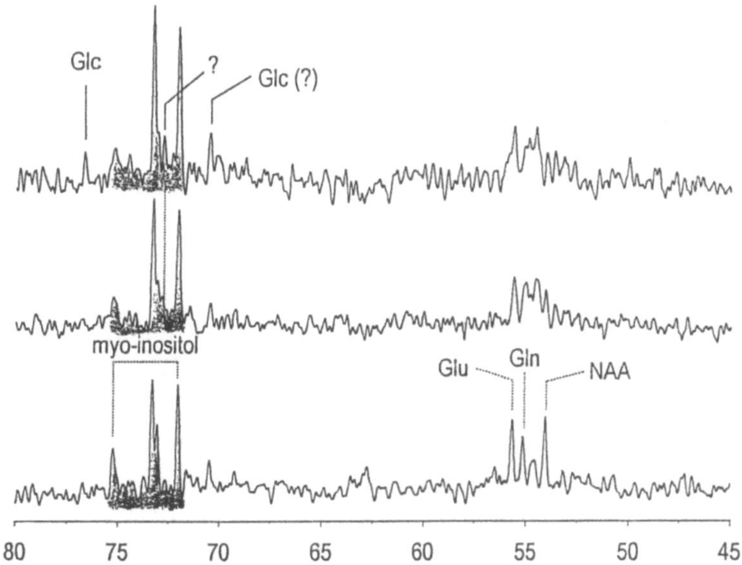

Fig. 6. 4T in vivo brain spectrum. Natural-abundance ^{13}C. Two patients with probable Alzheimer disease (AD; *top*, *center*) and one normal subject *(bottom)* are compared. Several abnormalities of AD brain are identified, including excess myoinositol and reduction of both *N*-acetyl aspartate (NAA) and glutamine. (From Gruetter et al. 1996)

5.3.4 Multinuclear Magnetic Resonance Spectroscopy

MRS has high specificity for metabolite assignment and detects several naturally occurring nuclei apart from ^{1}H and ^{31}P. In "natural abundance," ^{13}C accounts for only 1% of the total carbon but, by the use of the higher magnetic field MR scanners now becoming widely available, can be used to identify several crucial metabolites in health and disease (Fig. 6). As an example of the latter, natural-abundance ^{13}C myoinositol is elevated in human brain in patients with Alzheimer disease and can now be directly observed in a 12-min examination of the brain in vivo (Gruetter et al. 1996). Low natural abundance of ^{13}C, as of ^{15}N, lends itself to the applications of isotopic "tracer" techniques with MRS and

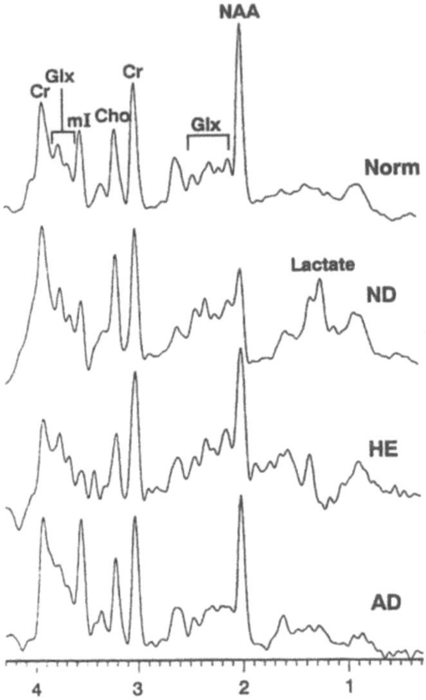

Fig. 7. Automated clinical ^{1}H MRS diagnosis and therapeutic monitoring. *mI*, myoinositol; *AD*, Alzheimer disease; *Cr*, creatine; *Glx*, glutamine plus glutamate;Cho, choline; *NAA*, *N*-acetyl aspartate; *ND*, near-drowning; *HE*, hepatic encephalopathy. (From Moats et al. 1995)

the identification of a very much broader range of natural metabolites. With ^{13}C MRS, for example, many of the intermediates of the tricarboxylic acid cycle (TCA) cycle and of glycolysis can be identified (see below), and with ^{15}N MRS, several crucial neurotransmitters can now be identified and their turnover in vivo monitored in real time (Kanamori and Ross 1997).

Table 1. Specificity of MRS for cell type

Metabolic function	Cell type or process	
^{1}H MRS	NAA	Neurons + axons
	Myoinositol	Astrocytes
	"Cholines"	Volume marker
^{31}P	Pi	pH
^{15}N	Glutamine	
	Glutamine synthase	Astrocytes
^{1}H-^{15}N	Glutamine amide proton	Astrocyte pH

5.4 Specificity

As shown in Fig. 7, automated, clinical ^{1}H MRS is a very simple means of recognizing commonly occurring patterns of neuropathology and is truly a metabolite imaging tool. It is obvious that MRS, however performed, averages the signals from several subpopulations of cells of widely different physiological and pathological significance. Despite this heterogeneity, MR signals specific to one or other subpopulation can sometimes be identified in the "averaged" spectrum. From this, some specificity of MRS for cell type can be deduced (Table 1).

While specificity in MR imaging (MRI) is heavily dependent upon contrast and contrast agents, and positron emission tomography (PET) has low chemical specificity unless combined with ligands and receptor probes, in vivo MRS is highly specific for both identity and concentration of selected metabolites.

In addition, although MRS is insensitive, and therefore has poor spatial resolution, it does have promising specificity for certain cell populations in a heterogeneous tissue.

5.5 Imaging of Receptor Expression

There are limited opportunities for directly imaging receptors with MRS. Few, if any, attemtps have been made to determine receptor number with MRS, principally because all calculations of the number and density of surface receptors indicate that they are likely to be below

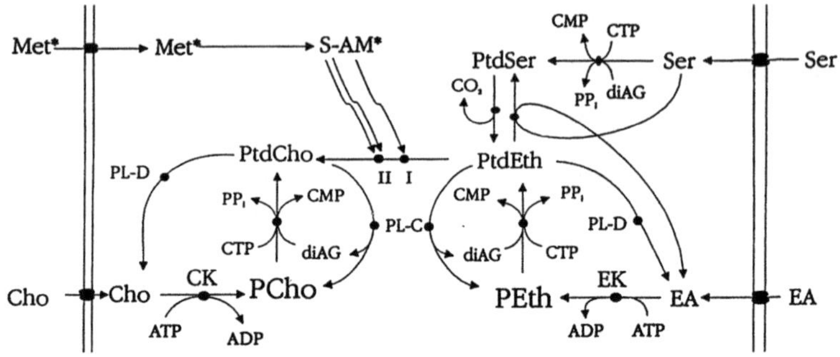

Fig. 8. Metabolism of membrane phospholipids. Major membrane consitutents of all cells include phosphatidyl-serine,-ethanolamine and -choline (*PtdSer; Ptd Eth* and *PtdCho*). Cho can be derived endogenously via the sequential S-adenylation of ethanolamine derivatives (abbreviated as EA, PEth and PtdEth in figure; for other abbreviations see original reference). (Reproduced with permission from Gillies et al. 1994)

the limit for direct detection by MRI. One approach, pioneered by Dr. Tom Meade of Caltech (Moats et al. 1997) involves engineering "smart" MR contrast agents. MR contrast agents use rare earths (e.g., gadolinium) or paramagnetic ions to alter the local magnetic enviornment, thereby altering relaxation of the intrinsic cell water. In Meade's approach, a molecule is synthesized which incorporates the contrast agent by a chemical bond which can only be broken by a specific intracellular enzyme. The molecule must be capable of entry into the cell intact; on reaching the enzyme-binding site, its breakdown is catalyzed, and the contrast agent that is now free becomes active. The result is currently designed to be "read" by MRI or MR microscopy, rather than by MRS, but it is probable that, were the contrast agent a chemical shift reagent rather than a relaxation agent, MRS might become the method of choice. Practical in vivo demonstrations of this exciting theoretical approach are awaited with interest.

A second approach is the identification of cell signaling molecules involved in apoptosis (cell death) (Hale et al. 1996). In Fig. 8 are shown a number of the intermediates of choline, ethanolamine, and inositol metabolism which can be directly assayed in vivo, using either ^{1}H or proton decoupled ^{31}P MRS. In gene therapies which impact cell death,

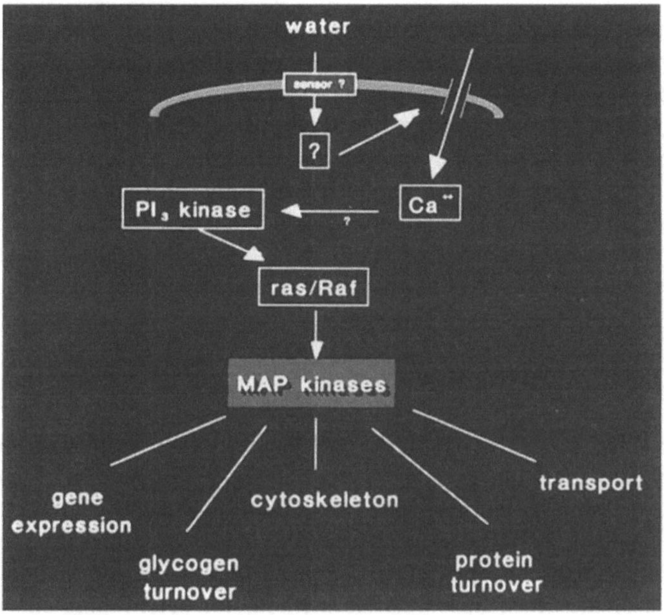

Fig. 9. Osmosignaling in astrocytes in vivo. *MAP*, mitogen-associated protein. (Unpublished figure, kindly provided by Prof. Dieter Haussinger)

for example in cancer therapy, it is recognized that some or all of these signals may alter, offering a means of noninvasive monitoring of the process.

The third approach to the problem of imaging alterations in changes in cell receptor expression arises from the recent recognition that cell volume is a potent signaling device in many tissues, including liver and brain. Several crucial kinases, especially mitogen-associated protein (MAP kinase), and oncogenes have been demonstrated to participate in the signaling cascade when the cell volume changes (Fig. 9). The signals, or osmolytes, are often measurable with MRS in vivo and include such ubiquitous metabolites as myoinositol, glycerophosphorylcholine, the amino acids glutamate and taurine, and possibly even *N*-acetylaspartate, the principal proton MRS marker of intact neurones.

Figure 9 indicates the state of knowledge of astrocyte osmoregulation available from in vivo MRS. Three significant clinical examples can be cited: hyponatremia (Videen et al. 1995), hypernatremia (Lee et al. 1994), and hepatic encephalopathy (Kreis et al. 1992).

In hepatic encephalopathy, in particular, disordered osmosignaling in the astrocyte appears to be implicated in some profound neuropsychological disturbance. By implication, accurate quantitative ^{1}H MRS in vivo offers some indirect means of "imaging" some of the processes and gene/enzyme expressions presented diagramatically in Fig. 9.

5.6 Enzyme Activity In Vivo

The principal method proposed for the demonstration of effective gene therapy is to determine in vivo enzyme activity, the prime result of new gene expression. In the direct demonstration and quantification of in vivo enzyme activity, MRS has been very successful to date, and here MRS is especially well placed to provide useful clinical information.

Each of the biochemist's classic methods for determining enzyme activity works well with MRS (Table 2). Some of the better known methods are as follows, with an illustrative example of each class given:

- Change in steady state concentration of substrate and/or product. *Example:* Glucose concentration in the brain of diabetics reflects enhanced transporter activity (Gruetter et al. 1992).
- Dynamic changes in concentration of intrinsic substrate(s) and product(s). *Examples:* Cardiac muscle response to stress (Weiss et al. 1990); skeletal muscle response to exercise; state of creatine kinase equilibrium after hypoxia. In this case, the spectra illustrated in Fig. 3 can be used to calculate free [ADP], ΔG, ATP, and CPK equilibrium after hypoxic (Bluml et al. 1996) or other neural stimuli, including activation in the visual cortex (Lee et al. 1995).
- Steady state concentration changes in product(s) after administration of substrate. *Examples:* Fructose load (effect on liver metabolites); ethanol load (appearance in brain) (Fig. 10).
- Instantaneous exchange between substrate and product in reversible enzyme reactions (Forsen and Hoffman 1963). *Example:* Creatine kinase flux and ATPase flux, determined by the transfer of magneti-

Table 2. Methods of determining enzyme activity in vivo

Methods	Stimulus/enzyme response
Loading tests	
Brain	Ammonia load
	Reye's syndrome
	Hepatic encephalopathy
	Glucose (also liver and muscle)
Liver	Fructose
Skeletal muscle	Exercise
Cardiac muscle	Stress
Enzyme flux at steady state	
^{13}C glucose	TCA cycle; transaminase
$^{15}NH_4Cl$	Glutaminase
	Glutamine synthetase
	Glutamate dehydrogenase
	Transaminases

TCA, tricarboxylic acid cycle.

zation between ATP and PCr or between ATP and inorganic phosphate, respectively (Mora et al. 1992).

Whereas these four methods are limited to the metabolites which are naturally abundant enough to reach detectable levels in MRS, the two that follow achieve greater flexibility and specificity by using a precursor which is enriched in MRS-detectable isotope(s) (currently, the favored methods employ ^{13}C, ^{15}N, or ^{19}F as MRS labels):

- Direct enzyme flux determination. *Example:* ^{13}C glucose flux into the TCA, a tool for assaying glucose transport, glutamate turnover, and TCA cycle rate in vivo (Mason et al. 1995). As illustrated, once it has entered the cell, the fate of glucose carbon can be followed by identification of the appropriate labeled resonance in the in vivo spectrum. Results have been very valuable in exploration of normal human brain and in application to the human liver and muscle in metabolic disease, principally diabetes mellitus. Essentially, any successful gene therapy involving one or more of the site-specific

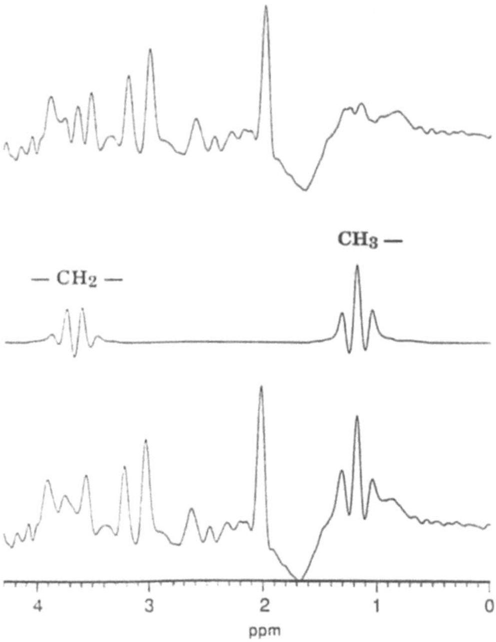

Fig. 10. Appearance of ethanol in human brain ^{1}H MRS. *Top*, difference spectrum; *center*, ethanol solution; *bottom*, alcoholic subject

steps illustrated in Fig. 11 would be demonstrable. ^{13}C MRS can now be performed on all high-field clinical MR scanners equipped with a second radiofrequency channel for decoupling.

^{13}C compounds are widely available, being synthesized de novo or in biological fermentations with ^{13}C nutrients, from which the desired substrate can be purified in large amounts. MR sensitivity limits are once again the issue, as the steady state concentration of the compound to be observed must be sufficient for its detection in a reasonable scan time, say 10–15 min. To date, most clinical investigations have confined themselves to starting with U^{13}C glucose. However, this limitation is largely temporary and will be overcome by cost reduction of the desired ^{13}C precursor and by higher field

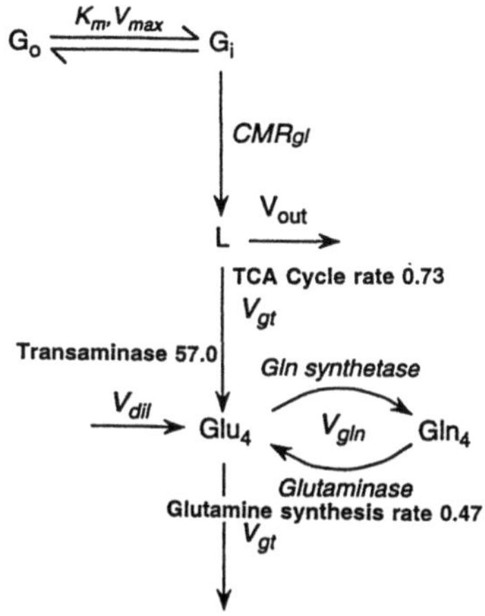

Fig. 11. ^{13}C flux determination in in vivo brain. After loading with intravenous ^{13}C-enriched glucose and infusion to establish steady state, brain ^{13}C MRS monitors changing concentrations of glutamate and glutamine C-4, from which the enzyme fluxes shown can be derived. *CMRgl*, cerebral metabolic rate of glucose; *TCA*, tricarboxylic acid cycle; *V*, velocity; *out*, lactate efflux; *dil*, dilu-ation by exogenous, unlabelled metabolites; *gln*, glutamine cycling; *gt*, glutamine synthetase. (From Mason et al. 1995)

strengths, which are now becoming available in clinical MR scanners and will improve sensitivity.

– Substrate chase method for measuring metabolic flux in vivo (Kanamori and Ross 1993). The general approach in direct enzyme flux determination necessitates a number of assumptions as the substrate moves through a long series of linked enzyme reactions to enter the TCA cycle. Modeling of the various steps, although complex has been generally successful. The substrate chase method is less dependent upon such complex models. Recent work with ^{15}N MRS (Fig. 12) from which in vivo glutaminase activity was determined (Kanamori and Ross 1995) serves as an example. Brain [5-^{15}N] glutamine can be detected with higher sensitivity by ^{1}H-^{15}N heteronuclear multiple-quantum coherence (HMQC) transfer MRS (Fig. 13) for measurement of in vivo glutamine synthetase activity at near-physiological state (Kanamori et al. 1996).

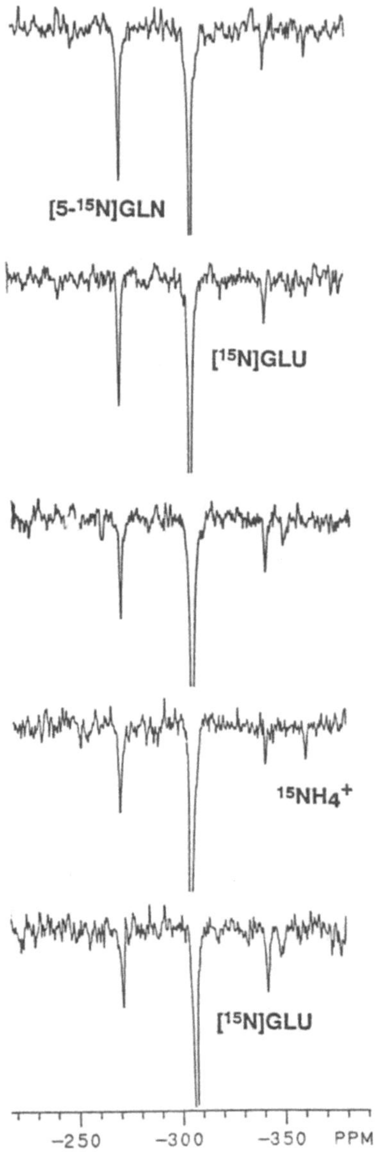

Fig. 12. Legend see p. 61

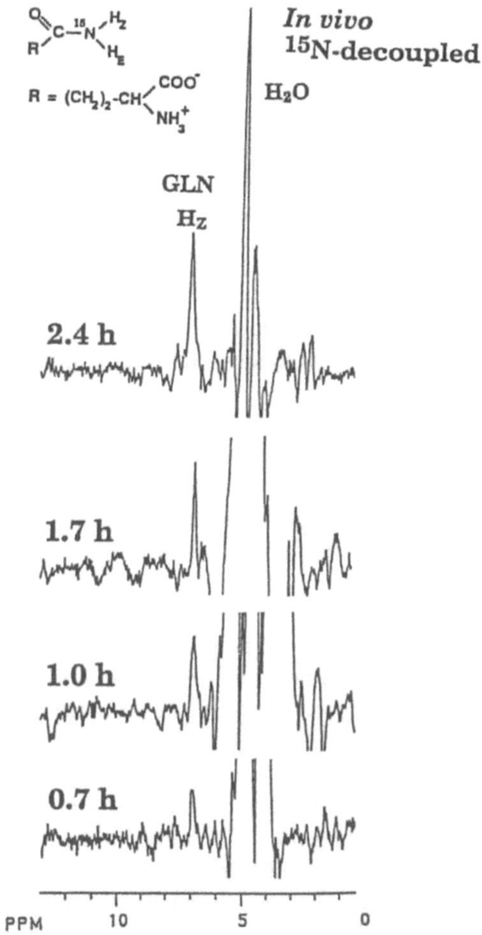

Fig. 13. Sequential in vivo ^{1}H-^{15}N MR brain spectra. Following infusion of $^{15}NH_4Cl$, the activity of in vivo glutamine synthetase was determined from the rate of increase in brain [5-^{15}N] glutamine amide proton H_Z

Fig. 12. Sequential in vivo ^{15}N magnetic resonance brain spectra infusion of ◀ $^{15}NH_4Cl$ and $^{14}NH_4Cl$ "chase." The glutaminase rate is determined from the reducing intensity of the 5-^{15}N-glutamine, from top to bottom

5.7 Conclusion

Enzyme activity in vivo can differ severalfold from that assayed in vitro by conventional means. This opens greater opportunities for MRS in testing emerging gene therapies and distinguishing between effective and ineffective protein–enzyme synthesis. Some of the tools which have been applied are listed below:

1. Imaging agents; chemical shift reagents, e.g., Europium; "smart" contrast agents
2. Cancer markers; lymphocyte studies
3. Gene deletion or replication models (transgenic mice; Downs syndrome in humans)
4. Gene replacement; bone marrow transplant
5. Animal models; other organs; liver, kidney, spleen transplant, and drug toxicity
6. Isolated, perfused organs liver; kidney; heart
7. Pharmacological research: fluorinated drugs; fluorinated metabolites

In vivo MRS in animal models can be performed without killing valuable mutant animals. Isolated perfused organs can be used to increase substrate concentrations beyond what may be achieved in vivo, thereby optimizing the MRS assay in a near-physiological setting.

One approach in cancer therapy has been to use specific cell markers in vitro (Mountford et al. 1990). The difficult task of monitoring these markers in vivo is just commencing. However, it is already apparent that some discoveries in MRS are changing the way in which molecular biologists view genetic markers. Some genes are recognized through tyrosine kinase, inositol phosphates, and other low-abundance markers. Others might be reflected in MR-visible metabolites. PME is the global term for a group of recently rediscovered membrane markers which are greatly altered in tumor cells of all kinds. This large family of molecules is likely to include at least one (i.e., ethanolamine) marker of a specific oncogene. In the fortunate event that this is so, in vivo MRS will have moved one step closer to becoming the clinical imaging tool of choice in the monitoring of experimental gene therapy.

Gene therapy, when successful, will result in significantly increased enzyme activity and altered local metabolite concentrations. Generally, MRS is too insensitive to determine the immediate changes in receptor activity or intracellular messengers involved in the genetic response. However, as a specific noninvasive metabolic tool, particularly adapted to determining enzyme activity in vivo, MRS may ultimately provide imaging methods of value in monitoring appropriate gene therapies.

Acknowledgements. B.D.R. is grateful to Schering Foundation for a travel grant to attend this symposium. The work performed was significantly assisted by Dr. David Dubowitz, Dr. Rex Moats, Tuan Huang, and Jeannie Tan and by financial support from Schulte Research Institute, Jameson Foundation, Whittier Family Foundation, and NIH Grant No. 1RO1 NS29048. The manuscript was skillfully prepared by Ms. Mary Munoz.

References

Bluml S, Ross B (1997) Clinical application of quantitative proton decoupled ^{31}P MRS in normal and diseased human brain. Proceedings of the 5th International Society for Magnetic Resonance in Medicine (abstr), vol 2, p 1171

Bluml S, Haseler LJ, Danielsen ER, Ross BD (1996) Creatine kinase and free-energy of ATP hydrolysis in normal and diseased human brain. Proceedings of the 4th Society of Magnetic Resonance, New York, vol 1, p 467

Bluml S, Zuckerman E, Tan J, Ross BD (1997) Osmotic and metabolic disturbances in human hepatic encephalopathy: a study with quantitative proton decoupled ^{31}P MRS. J Clin Invest (submitted)

Ernst T, Kreis R, Ross BD (1993) Absolute quantitation of water and metabolites in the human brain, part I: compartments and water. J Magn Reson 102 (1):1–8

Forsen S, Hoffman RA (1963) Study of moderately rapid chemical exchange reactions by means of nuclear magnetic double resonance. J Chem Phys 39 (11):2892–2901

Gillies R, Barry J, Ross B (1994) In vitro and in vivo ^{13}C and ^{31}P NMR analyses of phosphocholine metabolism in rat glioma cells. Magn Reson Med 32:310–318

Gruetter R, Novotny EJ, Boulware SD, Rothman DL, Mason GF, Shulman GI, Shulman RG, Tamborlane WV (1992) Direct measurement of brain glucose concentrations in humans by ^{13}C NMR. Proc Natl Acad Sci USA 89:1109

Gruetter R, Danielsen ER, Ross BD (1996) A ^{1}H and ^{13}C MRS study of Alzheimer disease at 4 Tesla. Proceedings of the European Society of Magnetic Resonance in Medicine and Biology, Prague, 1996

Hale AJ, Smith CA, Sutherland LC, Stoneman VE, Longthorne VL, Culhane AC, Williams GT (1996) Apoptosis: molecular regulation of cell death. Eur J Biochem 236:1–26

Kanamori K, Ross BD (1993) ^{15}N NMR measurement of the in vivo rate of glutamine synthesis and utilization at steady state in the brain of the hyperammonaemic rat. Biochem J 293 (2):461–468

Kanamori K, Ross BD (1995) In vivo activity of glutaminase in the brain of hyperammonaemic rats measured by ^{15}N nuclear magnetic resonance. Biochem J 305:329–336

Kanamori K, Ross BD (1997) In vivo nitrogen MRS studies of rat brain metabolism. In: Bachelard H (ed) Advances in neurochemistry. Plenum, New York (in press)

Kanamori K, Ross BD, Chung JC, Kuo EL (1996) Severity of hyperammonemic encephalopathy correlates with brain ammonia level and saturation of glutamine synthetase in vivo. J Neurochem 67:1584–1594

Koretsky A, Brosnan M, Chen L, Chen J, Van Dyke T (1989) Expression of creatine kinase in the livers of transgenic mice: in vivo detection by ^{31}P NMR (abstract). Soc Magn Reson Med 1:258

Kreis R, Ross BD, Farrow NA, Ackerman Z (1992) Metabolic disorders of the brain in chronic hepatic encephalopathy detected with ^{1}H MRS. Radiology 182:19–27

Lee JH, Arcinue E, Ross BD (1994) Organic osmolytes in the brain of an infant with hypernatremia. N Engl J Med 331 (7):439–442

Lee JH, Ross BD, Farrow N, Ernst T (1995) Visual stimulation alters redox state of human cerebral cortex. Proceedings of the 3rd Society of Magnetic Resonance, Nice, France, vol 2, p 784

Mason GF, Gruetter R, Rothman DL, Behar KL, Shulman RG, Novotny EJ (1995) Simultaneous determination of the rates of the TCA cycle, glucose utilization, alpha-ketoglutarate/glutamate exchange, and glutamine synthesis in human brain by NMR. J Cereb Blood Flow Metab 15:12–25

Moats RA, Watson L, Shonk T, Tokuyama S, Braslau D, Eto R, Mandigo JC, Ross BD (1995) Added value of automated clinical MRS of the brain. J. Comput Assist Tomogr 19:480–491

Moats R, Frazier S, Meade T (1997) "A smart" magnetic resonance imaging agent that reports on specific enzymatic activity. Angw Chem Int (Engl) (in press)

Mora BN, Narasimhan PT, Ross BD (1992) 31P magnetization transfer studies in the monkey brain. Magn Reson Med 26:100–115

Mountford CE, Delikatny EJ, Dyne M, Holmes KT, Mackinnon WB, Ford R, Hunter JC, Truskett ID, Russell P (1990) Uterine cervical punch biopsy specimens can be analyzed by ^{1}H MRS. Magn Reson Med 13:324–331

Ross BD, Michaelis T (1994) Clinical applications of magnetic resonance spectroscopy. Magn Reson Quart 10 (4):191–247

Videen JS, Michaelis T, Pinto P, Ross BD (1995) Human cerebral osmolytes during chronic hyponatremia: a proton magnetic resonance spectroscopy study. J Clin Invest 95:788–793

Webb PG, Sailasuta N, Kohler S, Raidy T, Moats RA, Hurd RE (1994) Automated single-voxel proton MRS: technical development and multisite verification. Magn Reson Med 31 (4):365–373

Weiss R, Bottomley P, Hardy C, Gerstenblith G (1990) Regional myocardial metabolism of high energy phosphates during isometric exercise in patients with coronary artery disease. N Engl J Med 323 (23):1593–1600

6 Tumour Targeting Using Peptides, Monoclonal Antibodies and Fragments

K.E. Britton

6.1 Introduction . . . 67
6.2 Antigens and Monoclonal Antibodies . . . 68
6.3 Radiolabel . . . 69
6.4 Use in Cancer . . . 70
6.4.1 Ovarian and Breast Cancer . . . 70
6.4.2 Colorectal Cancer . . . 70
6.4.3 Prostate Cancer . . . 72
6.4.4 Miscellaneous Tumours . . . 72
6.5 Radionuclide Receptor Imaging . . . 73
6.6 New Two- and Three-Stage Imaging . . . 74
6.7 Genetic Engineering . . . 76
6.8 Conclusion . . . 77
References . . . 77

6.1 Introduction

The cell surface is a complex reflection of the events in its nucleus and cytoplasm. At the same time, the activities of the nucleus and the cytoplasm depend on chemical information transmitted from the cell surface by signal transduction. This interaction results in exploitable differences between the surface properties of a cancer cell and its normal and benign counterparts. Tumour targetting using radiolabelled peptides, monoclonal antibodies and fragments makes use of all of these properties.

The requirements for these new imaging techniques are that they answer clinically relevant questions that cannot be answered by conventional or advanced radiological techniques. These include the demonstration of subclinical and subradiological disease, the detection of recurrences before elevation of serum markers, the location of the site of recurrence when serum markers are elevated, especially when radiology is negative, the demonstration that a clinically or radiological evident mass is viable tumour and not just post-surgical or post-therapy fibrosis and the evaluation of the effect of chemotherapy without the need for a "second look" operation.

6.2 Antigens and Monoclonal Antibodies

The requirements for radioimmunoscintigraphy include the following:

1. Selection of an antigen as specific as possible to the cancer and a monoclonal antibody directed against this antigen
2. The best radiolabel appropriate to the kinetics of uptake of the antibody by the cancer
3. A radiolabelling method that preserves immunoreactivity
4. A gamma camera imaging system optimised for the radiolabel and the site of cancer
5. An analysis system that makes best use of imaging data

Tumour-associated antigens come in a number of varieties. Epithelial surface antigens are separated from the blood by being in the inner surfaces of the cells and by biological barriers, but exposed to blood by the architectural disruption of the malignant process. Such antibodies include HMFG1, HMFG2 and SM3 for ovarian (Granowska et al. 1993a) and breast cancer (Granowska et al. 1996) and PR1A3 for colorectal cancer (Granowska et al. 1990a). A second well-used group are oncofetal antigens such as carcinoembryonic antigen (CEA; Baum et al. 1989). Antibodies against tumour-derived antigens such as MOV18 (Crippa et al. 1991) and B72.3 (Collier et al. 1993) are well known. Antibodies against viral antigens may be used for imaging tumours such as hepatoma, and ones against synthetic antigens for certain other tumours, e.g. 170H82 antibody (Alexander et al. 1995).

6.3 Radiolabel

The choice of radiolabel has evolved from the use of ^{131}I, which is poor, to ^{123}I, which is difficult and expensive to obtain but chemically useful, to ^{111}In to ^{99m}Tc. The disadvantages of ^{111}In need to be appreciated. It has to be ordered and it is the most expensive. The two-step labelling method has to be undertaken with metal-free reagents. It gives a relatively high dose of absorbed radiation, typically an effective dose of around 24 mSv for a normal injection of 120 MBq. A medium energy collimator is required. There is high bone marrow and liver uptake through deposition of the indium metal in the reticulo-endothelial system, and there is often abnormally high large-bowel activity. Conversely, ^{99m}Tc is readily available routinely at low cost and there are rapid labelling techniques with complete in vivo stability. Thyroid blockade is not required. The high count rate, short-lived radionuclide gives high-quality images and is also suitable for single photon emission computed tomography (SPECT). The study can be arranged and completed within 24 h, and there is low absorbed radiation (effective dose about 3 mSv for 750 MBq). There is low liver, marrow and bowel uptake compared with ^{111}In.

It is the signal that matters, not the signal-to-noise ratio. Much of the noise in the image depends on the square root of the signal count rate. The lower the count rate, the proportionally larger the error on the signal and the proportionally larger the error on the background (Britton and Granowska 1987). Antibody uptake is maximal between 6 and 24 h; therefore, ^{99m}Tc, ^{123}I and ^{111}In are appropriate radiolabels. The shorter the half-life, the greater the administered activity that can be given, the higher the count rate achieved, the lower the noise due to the signal, the earlier the detection and the smaller the lesion detected. The fact that the signal is so strong when imaging with antibodies, fragments or peptides is due to the high amplification factor. Each cancer cell may have five to 30000 binding sites for the agent, of which about one out of 50 is radiolabelled. The affinity for binding ranges from 10^8 to 10^{14} l/mol, and the residence time may be days or weeks on the tumour.

6.4 Use in Cancer

6.4.1 Ovarian and Breast Cancer

There are a range of antibodies which can be used for ovarian cancer. Examples include the use of ^{123}I-labelled HMFG2 (Epenetos et al. 1982; Granowska et al. 1984, 1986), ^{111}In-labelled HMFG2 (Granowka et al. 1991), ^{99m}Tc-labelled HMFG1 and ^{99m}Tc-labelled SM3 (Granowka et al. 1990b, 1993) in imaging both primary and recurrent ovarian cancer. The technique is particularly able to demonstrate the presence of remaining disease after chemotherapy and to avoid the need for a "second look" operation. This technique can be combined with kinetic analysis and probability mapping, and tumour recurrences down to half a centimetre can be demonstrated (Granowka et al. 1988).

Another application for ovarian cancer imaging is after screening well women or women with a high familial risk for ovarian cancer using ultrasound. Even with transvaginal ultrasound and colour Doppler, the ultrasound techniques are not specific to cancer. Using the SM3 monoclonal antibody, which has an average malignant to benign uptake ratio of 17:1 (compared with an average ratio of 6:1 for HMFG2), should enable a clear distinction to be made between malignant causes of pelvic masses discovered by ultrasound and those that are not malignant. Borderline malignant tumours also express SM3 antigen. The use of ^{99m}Tc-SM3 has given a level of accuracy of the order of 93% (Granowka et al. 1993a). The same antibodies labelled with ^{99m}Tc may be used for breast cancer, particularly SM3 in the detection of involvement in clinically impalpable nodes (Granowka et al. 1996). The monoclonal antibody H170 against a synthetic glycoprotein antigen is also successful (McEwan et al. 1994).

6.4.2 Colorectal Cancer

A wide range of monoclonal antibodies may be used, such as ^{111}In-labelled CEA (Granowka et al. 1984, 1989) and ^{111}In-labelled PR1A3 (Granowska et al. 1991) or ^{99m}Tc-labelled PR1A3 (Granowka et al. 1993b). If an antigen is not fixed to the tumour but appears in the lymphatic system, then nodal involvement will be misdiagnosed because of the presence of the antigen rather than the presence of malignant cells when imaging. This is true for CEA and B72.3 antigens and has been particularly emphasised when intra-operative probe work is done in which false localisation of tumour in lymph nodes may be obtained (Kuhn et al. 1991). A 10-min image should be taken to act as a template with which to compare the later images. Non-specific uptake has a high initial distribution seen on the early blood pool image, which then decreases with time, whereas specific uptake increases with time. These kinetic factors are very important aids in distinguishing between malignant disease and other diseases. They also allow the determination of very small disease sites by comparing early and late images. Even visually, tumours down to 1.26 g have been demonstrated. By change detection analysis, axillary lymph nodes down to 0.35 g that are involved in breast cancer have been correctly identified by imaging (Granowska et al. 1996).

It is required that the antibody should react with all the relevant tumours in the class. This is a problem with B72.3, which does not react with up to 20% of colorectal cancers; thus a sensitivity rate of below 80% is inherent in the technique. It is important that the antibody reacts with both poorly differentiated and well differentiated cancers. For example, PR1A3 reacts three times better than anti-CEA with poorly differentiated colorectal cancer (Granowska et al. 1989).

Using ^{99m}Tc-labelled PR1A3, a level of accuracy in primary tumour imaging of 97%, in recurrent abdominal and pelvic tumour of 96% and in detecting liver involvement of 84% was shown (Granowska et al. 1993b), with an overall accuracy of 92% in 131 studies. This compares with the use of ^{99m}Tc–labelled anti-CEA in colorectal recurrence, which has a level of accuracy of 86% in the liver, 95% in the abdomen and 77% in the pelvis (Lind et al. 1991), and of ^{111}In-labelled B72.3, which has a sensitivity rate of 73% (58 out of 79 patients) in the abdomen and pelvis (Doerr et al. 1991). Computed tomography (CT) scanning in the same series gave a sensitivity rate of 53% (42 out of 79).

The per-operative probe is a useful adjunct to the imaging of colorectal cancer. The original work carried out by Martin used ^{125}I-labelled antibodies, but a delay of up to 30 days was required between the injection of the radiolabelled antibody and the probe study (Martin et al. 1985). This can be shortened using a two-stage approach (Paganelli et al. 1994). ^{99m}Tc antibodies may be used with an appropriate probe after completion of the imaging (Howell et al. 1995).

6.4.3 Prostate Cancer

There is a 1 in 7 chance of prostate cancer developing in a male aged over 55. Earlier studies with anti-prostate-specific antigen (anti-PSA) antibodies were initially successful, but failed when the PSA level in the blood was high because the circulating PSA bound the antibody given for imaging, transferring the immune complex to the liver. Phase 1, 2 and 3 studies have been successfully undertaken using ^{111}In-labelled CYT356 (Cytogen Corp.) in the United States (Tyson et al. 1991; Babaian et al. 1994). This monoclonal antibody is against a prostate-specific membrane antigen. Imaging can be put to use in defining the preoperative extent of prostate cancer before radical prostatectomy. It is useful in demonstrating soft tissue recurrence, especially locally in situations where the serum PSA is rising, the bone scan is normal and ultrasound and/or CT scan are normal or equivocal. The patient protocol is the same as for other ^{111}In-labelled agents. ^{99m}Tc radioimmunoscintigraphy requires images at 10 min, 4–6 h and 24 h. SPECT, however, is essential at two time periods, such as 4 h and 24 h, because the increasing uptake of the antibody by the prostate cancer in the pelvis is best demonstrated by comparison of serial SPECT studies. Bone uptake occurs if the bone metastases are recent, but not if they are long-standing and have been treated with hormone therapy. This approach to prostate cancer imaging with the same antibody labelled with ^{99m}Tc is being further evaluated (Feneley et al. 1996).

6.4.4 Miscellaneous Tumours

^{99m}Tc-labelled monoclonal antibodies have been successfully used in the European multicentre trial and Italian studies of melanoma using the Sorin antibody 225.28S (Buraggi et al. 1985; Siccardi et al. 1986, 1990). $F(ab^1)_2/Fab^1$ mixture gives a higher rate of accuracy than the equivalent ^{111}In- and ^{131}I-labelled antibodies. It has also been used successfully in ocular melanoma with a rate of accuracy of 94% (Bomanji et al. 1987a).

Sheritz et al. (1994) have shown excellent imaging with ^{99m}Tc Fab[1] fragment LL2 in non-Hodgkin's lymphoma.

6.5 Radionuclide Receptor Imaging

A receptor accepts a chemical agent such as a hormone, a neurotransmitter, a cytokine or other biologically active peptides. The binding process initiates a train of reactions inside the cell so that information is moved chemically from the cell surface to the nucleus. This is called signal transduction. The first step typically involves a G-protein (Fraser 1994), which is a complex molecule fixed in the cell membrane with a part external to the membrane to bind and a part internal to the membrane to initiate or mediate chemical activity in the cytoplasm through its conformational change once binding has occurred. A typical chemical activity is the activation of cyclic AMP and then tyrosine phosphokinases. The chain of reactions leads to the nucleus and binding of activators or suppressors to particular genes. In this way, the synthesis of control proteins is either enhanced or suppressed. The cell cycle may be set in motion or differentiation may occur. Inhibitory hormones such as somatostatin, a growth hormone release-inhibiting hormone, acts in this way. It has a synthetic analogue octreotide (Ur et al. 1993a,b; Krenning et al. 1993), which has been labelled with ^{123}I or ^{111}In and called pentreotide (Krenning et al. 1993, 1995; Lamberts et al. 1991). Stimulatory agents such as α-melanocyte-stimulating hormone (α-MSH) have been labelled with ^{123}I for imaging melanoma (Bard et al. 1992), and vasoactive intestinal peptide (VIP) labelled with ^{123}I for imaging gastrointestinal cancers (Virgolini et al. 1994). The normal routine neurotransmitter analogue metaiodobenzylguanidine (MIBG) labelled with ^{123}I for imaging or ^{131}I for therapy is in routine use in patients with neural crest tumours and neuroblastoma (Bomanji et al. 1987b; Shapiro et al. 1985; Hofnagel et al. 1994). The cytokine interleukin (IL)-2 has been labelled with ^{123}I or ^{99m}Tc and used to image activated T cells in autoimmune diseases (Barone et al. 1996) and T cell responses to tumours such as renal cell carcinoma and melanoma (Signore et al. 1994). There are at least another 100 biologically active peptides to explore. Indeed, as the binding sites of antibodies are peptides, so there is a convergence of both fields to the development of ^{99m}Tc peptides. The requirement is for a cancer-specific binder generously expressed on the cancer cell surface and a ^{99m}Tc-labelled or β-radionuclide therapy bindee of high affinity, stability and residence time to target the specific cell binder (Britton and Granowka 1993; Britton 1995).

Finally, the nuclear binding of sense and antisense oligonucleotides in DNA and RNA are being explored (Hnatowich et al. 1995; Urbain et al. 1995). ^{111}In and ^{99m}Tc oligomers of 15–22 base pairs (mers) have been synthesised, raising the possibility of direct imaging of oncogenes or display of the selectivity of gene therapy.

6.6 New Two- and Three-Stage Imaging

The merits of imaging with radiolabelled antibodies have often been ignored by commercial companies as they seek to use the same technique to target radionuclide-labelled antibodies to treat cancer. This is much more difficult because of three main reasons. Firstly, for imaging it does not matter if only the outer shell of the tumour takes up the antibody; the signal from the tumour will be evident. For therapy, the antibody with its therapy label must target the middle of the tumour. Secondly, the amount of antibody and its residence time on the tumour can be quite small for imaging, whereas for therapy it must bring a reasonable amount of the radionuclide to irradiate the tumour over a prolonged period. Thirdly, and most importantly, for imaging it does not matter that the normal tissues are bathed in radiolabelled antibody since the radiation dose is very small, whereas for therapy the irradiation of normal tissues by radionuclide-labelled antibody not taken up by tumour (about 97% of the administered antibody) is potentially harmful to the bone marrow and gives a poor therapeutic ratio.

In order to overcome this problem, a series of clever strategies have been devised involving two or three stages of targetting the tumour. The most widely tested is the avidin–biotin system (Paganelli et al. 1993).

Biotin is a small molecule (vitamin B_6) which binds very strongly to the protein avidin at four different sites. Biotin is attached to the monoclonal antibody chosen to bind to the cancer, and this non-radioactive combination is injected first.

Avidin is injected 2 or 3 days later after the antibody has been metabolised in the blood and normal tissues, but not on the tumour.

Avidin therefore binds to the biotin attached to the antibody attached to the tumour. Non-bound avidin clears any unbound biotin-linked antibody into the liver and reticulo-endothelial system. One day later, radiolabelled biotin is injected. For imaging, the radiolabel may be ^{111}In or ^{99m}Tc; for therapy, it may be ^{90}Y. The third stage with the biotin-linked radionuclide targets the avidin which is bound to the biotin to the antibody to the tumour. Unbound biotin-linked radionuclide rapidly clears through the kidneys, so the irradiation of normal tissues is short, whereas the residence time of the radiolabelled biotin attached to the tumour is long, giving a good therapeutic ratio. Variations on this theme for imaging include ^{111}In-labelled avidin with biotin-linked antibody injected first or ^{99m}Tc-labelled biotin with avidinated antibody injected first.

An alternative approach is to use a bifunctional antibody. These can be created by linking two Fab or Fv groups together, one group binding to the cancer and the other to a radiotherapy-labelled molecule called a ligand. The linkage between the two fragments may be chemical or through genetic engineering, a diabody (Holliger et al. 1993). The bifunctional antibody (non-radioactive) is injected first and binds the tumour. Unbound bifunctional antibody is cleared into the liver and reticulo-endothelial system. Two or 3 days later, the radiolabelled ligand is injected and picks out the bifunctional antibody bound to tumour. Unbound radiolabelled ligand is rapidly excreted through the kidneys, again giving a good therapeutic ratio (Britton et al. 1993).

A further variant on this approach is to use an oligonucleotide that is a small molecule made up of base pairs in the same way as DNA. Since DNA has the ability to replicate, it has "sense" and "antisense" complimentary chains. For the two-step approach, a sense oligonucleotide is attached to the monoclonal antibody designed to target the cancer. The radiolabel is attached to an antisense oligonucleotide which will only bind with the sense oligonucleotide attached to the tumour, thereby targeting the radiolabel for imaging or therapy. These developments in the field are tested out by using imaging radionuclides, but are designed for radioimmunotherapy (Hnatowich et al. 1995; Urbain et al. 1995).

6.7 Genetic Engineering

The formation of all proteins by cells is controlled by genes, and antibodies are no exception. It is possible to isolate the genes that produce the heavy chains and the genes that produce the light chains and to combine them in a yeast or in a bacterial form called a plasmid. The plasmid will then produce both the chosen antibody and its own proteins from its genetic intruder and its own genes. This is similar to the way that a virus infects the cell and injects its own DNA or RNA into the host genetic DNA or RNA and therefore replicates itself and its own instructions to build itself again. In fact, bacteriophages can be used to aid this genetic transformation of the host plasmid. This ability to transfer human fragments of genes has been helped by the development of the polymerase chain reaction (PCR), which is a method of replicating DNA or RNA sequences millions of times over using a particular enzyme polymerase. This genetic material in plasmids may be combined together in a random fashion to make libaries of heavy and light chain-synthesising systems. In this way, it is hoped that new monoclonal antibody-like molecules will be developed with quite different binding properties than can be found in nature (Winter et al. 1994; Chester et al. 1994). Moreover, by using human material, the human anti-mouse response can be avoided. This response is not a problem for imaging, but it is a problem when large amounts of mouse monoclonal antibody are used in therapy.

At present, chimeric antibodies are available in which the Fc portion of the mouse antibody is replaced by a human Fc; in addition, human reshaped antibodies, also called complementarity-determining region (CDR) -grafted antibodies, are also available in which both the Fc portion and the framework of the mouse antibody are replaced by human antibody, leaving only the CDRs to be of murine origin. There are further advantages for genetically engineered molecules. For example, a sequence that will bind ^{99m}Tc can be introduced into the peptide; alternatively, the binding affinity may be enhanced by changing one amino acid for another. Thus quite novel peptides can be produced which are based on, but no longer are, monoclonal antibodies. They have more in common with receptor-binding peptides such as ^{111}In octreotide. Indeed, there is a convergence between the receptor peptides that are being developed and synthesised and the monoclonal antibody-derived genetically engineered peptide molecules that are being created. All are examples of the binder–bindee concept where a bindee is chosen for radiolabelling and to react with a specific binder property on the tumour cell surface.

The field is progressing further with the use of oligonucleotides themselves as imaging and therapy agents. In this case, they are made resistant to destructive enzymes in the plasma, e.g., as thiophosphates (Hnatowich et al. 1995). They can enter the cell and the cell nucleus. If the correct antisense sequence is chosen, they can bind directly with the particular abnormal onocogene sequence present in the cancer cell. The potential for imaging and therapy in these ways is enormous.

6.8 Conclusion

Radioimmunoscintigraphy and radiopeptide scintigraphy are robust imaging techniques whose time has come. Their main use is the identification of cancer that cannot be detected clinically, radiologically or by serum markers. Developments of monoclonal antibodies and peptides for imaging and therapy of virtually every type of cancer are being researched in academic departments. Their applications are mainly limited not by science, but by the hostile regulatory environment that prevents their rapid and cost-effective commercial development.

References

Alexander C, Villena-Heinsen CE, Trampert K et al (1995) Radioimmunoscintigraphy of ovarian tumours with technetium-99m labelled monoclonal antibody-170: first clinical results. J Nucl Med 22:645–651

Babaian RJ, Sayer J, Podoloff DAS et al (1994) Radioimmunoscintigraphy of pelvic lymph nodes with ^{111}indium-labelled monoclonal antibody CYT-356. J Urol 152:1952–1955

Bard DR, Knight CG, Wraight EP et al (1992) The use of a chelating derivative of alpha melanocyte stimulating hormone for the clinical imaging of malignant melanoma. Br J Radiol 65:112–118

Barone B, Chianelli M, Bottoni U et al (1996) In vivo detection of lymphocytic infiltration in melanoma using 99m-Tc-IL2. Eur J Nucl Med 23:1217

Baum RP, Hertel A, Lorenz et al (1989) ^{99m}Tc-labelled anti-CEA monoclonal antibody for tumour immunoscintigraphy: first clinical results. Nucl Med Commun 10:345–348

Bomanji J, Hungerford JL, Granowska M, Britton KE (1987a) Radioimmunoscintigraphy of ocular melanoma with ^{99m}Tc labelled cutaneous melanoma antibodies. Br J Ophtalmol 71:651–658

Bomanji JB, Levison DA, Flatman WD et al (1987b) Uptake of I-123 metaiodobenzylguanidine by phaeochromocytomas other paragangliomas and neuroblastomas: a histopathological comparison. J Nucl Med 28:973–978

Britton KE (1995) The tissue characterisation of malignancy with radionuclide studies. In: Bergmann H, Sinzinger M (eds) Radioactive isotopes in clinical medicine and research. 21st Bad Gastein symposium. Birkhauser, Basel, pp 249–280

Britton KE, Granowska M (1987) Radioimmunoscintigraphy in tumour identification. Cancer Surv 6:247–267

Britton KE, Granowska M (1993) The present and future of radiolabelled antibodies in oncology. Ann Nucl Med 7:127–132

Britton KE, Mather SJ, Granowska M (1993) Radiolabelled monoclonal antibodies in oncology. III. Radioimmunotherapy. Nucl Med Commun 12:333–347

Buraggi GL, Callegaro I, Mariani G et al (1985) Imaging with 131-I-labelled monoclonal antibodies to a high-molecular weight melanoma-associated antigen in patients with melanoma: efficacy of whole immunoglobulin and its $F(ab^1)_2$ fragments. Cancer Res 45:3378–3387

Chester KA, Begent RH, Robson L et al (1994) Phage libraries for generation of clinically useful antibodies. Lancet 343:455–456

Collier BD, Abdel-Nabi H, Doerr RJ et al (1993) Radioimmunoscintigraphy performed with In-111-labelled-CYT-103 in the management of colorectal cancer: comparison with CT. Radiology 185:179–186

Crippa F, Buraggi GL, Di Re E et al (1991) Immunoscintigraphy of ovarian cancer with the MOV18 monoclonal antibody. Eur J Cancer 27:724–729

Doerr RJ, Abdel-Nabi H, Krag D, Mitchell E (1991) Radiolabelled antibody imaging in the management of colorectal cancer. Arch Surg 214:118–124

Epenetos AA, Britton KE, Mather SJ et al (1982) Targetting of iodine-123-labelled tumour associated monoclonal antibodies to ovarian, breast and gastrointestinal tumours. Lancet 2:999–1004

Feneley M, Chengazi VU, Kirby RS et al (1996) Prostatic Radioimmunoscintigraphy: preliminary results using Tc-99m labelled monoclonal antibody, CYT351. Br J Urol 77:373–381

Fraser CM (1994) Structure and functional analysis of G protein-coupled receptors and potential diagnostic ligands. J Nucl Med 36 [Suppl]:175–215

Granowska M, Shepherd J, Britton KE et al (1984) Ovarian cancer: diagnosis using I^{123} monoclonal antibody in comparison with surgical findings. Nucl Med Commun 5:485–499

Granowska M, Britton KE, Shepherd JH et al (1986) A prospective study of 123-I labelled monoclonal antibody imaging in ovarian cancer. J Clin Oncol 4:730–746

Granowska M, Nimmon CC, Britton KE et al (1988) Kinetic analysis and probability mapping applied to the detection of ovarian cancer by radioimmunoscintigraphy. J Nucl Med 29:599–607

Granowska M, Jass JR, Britton KE, Northover MSA (1989) A prospective study of the use of ^{111}In-labelled monoclonal antibody against carcino-embryonic antigen in colorectal cancer and of some biological factors affecting its uptake. Int J Colorect Dis 4:97–108

Granowska M, Mather SJ, Britton KE et al (1990a) ^{99m}Tc radioimmunoscintigraphy of colorectal cancer. Br J Cancer 62 [Suppl X]:30–33

Granowska M, Mather SJ, Jobling T et al (1990b) Radiolabelled stripped mucin SM3 monoclonal antibody for immunoscintigraphy of ovarian tumours. Int J Biol Mark 5:89–96

Granowska M, Mather SJ, Britton KE (1991) Diagnostic evaluation of ^{111}In and ^{99m}Tc radiolabelled monoclonal antibodies in ovarian and colorectal cancer: correlations with surgery. Nucl Med Biol 18:413–424

Granowska M, Britton KE, Mather SJ et al (1993a) Radioimmunoscintigraphy with technetium-99m-labelled monoclonal antibody, SM3, in gynaecological cancer. Eur J Nucl Med 20:483–489

Granowska M, Britton KE, Mather SJ et al (1993b) Radioimmunosintigraphy with technetium-99m labelled monoclonal antibody, 1A3, in colorectal cancer. Eur J Nucl Med 20:690–698

Granowska M, Biassoni L, Carroll MJ et al (1996) Breast cancer Tc-99m SM3 radioimmunoscintigraphy. Acta Oncol 35:319–321

Hnatowich SJ, Mardirossian G, Fogartasi M (1995) Properties of single-stranded phosphodiester and phosphothioate DNA labelled with technetium-99m. J Nucl Med 36:103

Hofnagel CA, de Kraker J, Valdes Olmes RA, Voute PA (1994) 131-I-MIBG as a first line treatment in high risk neuroblastoma patients. Nucl Med Commun 15:712–717

Holliger P, Prosperio T, Winter G (1993) 'Diabodies' small bivalent and bispecific antibody fragments. Proc Natl Acad Sci USA 90:6444–6448

Howell R, Morris G, Granowska M et al (1995) Peroperative radioimmunodetection (PROD) of colorectal cancer using Tc-99m PR1A3 monoclonal antibody. Tumori 81 [Suppl]:107–108

Krenning EP, Kwekkeboom DJ, Bakker WH et al (1993) Somatostatin receptor scintigraphy with 111-In-DTPA-D-PheI- and 123-I-Tyr 3-octreotide: the Rotterdam experience with more than 1000 patients. Eur J Nucl Med 20:716–731

Krenning EP, Kwekkeboom DJ, Pauwels S et al (1995) Somatostatin receptor scintigraphy. Nucl Med Ann 1–50

Kuhn JA, Corbisiero RM, Buras RR et al (1991) Intraperoperative gamma detection with presurgical imaging in colon cancer. Arch Surg 126:1398–1403

Lamberts SWJ, Krenning EP, Reubi JC (1991) The role of somatostatin and its analogs in the diagnosis and treatment of tumours. Endocr Rev 12:450–482

Lind P, Lechner P, Adrian-Schad K et al (1991) Anticarcinoembryonic antigen immunoscintigraphy (technetium-99m monoclonal antibody BW 431/26) and serum CEA levels in patients with suspected primary and recurrent colorectal carcinoma. J Nucl Med 32:1319–1325

Martin EW, Hinkle GH, Tuttle S et al (1985) Intraoperative radioimmunodetection of colorectal tumour with a hand held radiation detector. Am J Surg 150:672–675

McEwan AJB, Maclean G, Goldberg L et al (1994) Evaluating Radioimmunoscintigraphy in patients with breast cancer. Eur J Nucl Med 21:748

Paganelli G, Malcovati M, Fazio F (1993) Monoclonal antibody pretargetting techniques for tumour localisation; the avidin biotin system. Nucl Med Commun 12:211–234

Paganelli G, Stella M, Zito F et al (1994) Radioimmunoguided surgery using iodine-125-labelled-18-biotinylated monoclonal antibodies and cold avidin. J Nucl Med 35:1970–1975

Shapiro B, Copp Je, Sisson JC et al (1985) I-131 metaiodobenzylguanidine for the locating of suspected phaeochromocytomas: experience in 400 cases. J Nucl Med 26:576–585

Sheritz J, Sharkey RM, Goldenberg DM (1994) Preclinical evaluation of a chimeric and humanised anti-B-cell lymphoma antibody. J Nucl Med 35:236

Siccardi AG, Buraggi GL, Callegaro L et al (1986) Multicentre study of immunoscintigraphy with radiolabelled monoclonal antibodies in patients with melanoma. Cancer Res 46:4817–4822

Siccardi AG, Buraggi GL, Natali PG et al (1990) European multicentre study on melanoma immunoscintigraphy by means of ^{99m}Tc-labelled monoclonal antibody fragments. Eur J Nucl Med 16:317–323

Signore A, Chianelli M, Ferretti E et al (1994) A new approach for the in vivo detection of insulinitis in type I diabetes: activated lymphocyte targetting with I-123 labelled interleukin-2. Eur J Endocrinol 131:431–437

Tyson IB, Heal AV, Tenorio L et al (1991) Immunoscintigraphy and pharmacokinetics of an ^{111}-In-Moab-C5.3 (CYT-356) for prostate cancer. J Nucl Med 32:1054

Ur E, Mather SJ, Bomanji J, Ellison D, Britton KE, Grossman AB (1993a) Pituitary imaging using a labelled somatostatin analogue in acromegaly. Clin Endocrinol (Oxf) 36:147–150
Ur E, Bomanji J, Mather SJ, Britton KE, Wass JAH, Grossman AB, Besser GM (1993b) Localization of neuroendocrine tumours and insulinomas using radiolabelled somatostatin analogies, ^{123}I-Tyr^{3}-octreotide and ^{111}In-pentetreotide. Clin Endocrinol (Oxf) 38:501–506
Urbain JLC, Shore SK, Vekmans MC et al (1995) Scintigraphic imaging of oncogenes with antisense probes: does it make sense? Eur J Nucl Med 22:499–504
Virgolini I, Raderer M, Kurtaran A et al (1994) Vasoactive intestinal peptide receptor imaging for the localisation of intestinal adenocarcinomas and endocrine tumours. N Engl J Med 331:1116–1121
Winter G, Griffiths AD, Hawkins RE, Hoogenboom HR (1994) Making antibodies by phage display technology. Annu Rev Immunol 12:433–455

7 Future Challenges in the Use of Positron Emission Tomography for In Vivo Molecular Imaging

T. Jones

7.1 Introduction 83
7.2 The Imaging Science of PET 85
7.3 Link with Molecular/Cell Biology and Drug Discovery and Development 87
7.4 The Challenge to Effect Translation 89
References 90

7.1 Introduction

The combined use of positron-emitting tracers and positron emission tomography (PET) provides the most specific and sensitive means for imaging molecular pathways and molecular interactions in vivo. Compared to other forms of medical imaging such as X-ray computed tomography (CT), magnetic resonance imaging (MRI) and ultrasound, PET is greater than 10^9 times more sensitive for measuring a tracer molecule's concentration in tissue, allowing sub-picomolar levels to be measured (Jones 1996a). This enables molecular pathways and molecular interactions to be studied with tracer doses of probe molecules without perturbing biological functions. In this regard, PET is seen as having a long-term future in the study of receptors, neurotransmitters, drug binding and general pharmacology.

Appreciable facilities and resources are needed to fully implement PET; hence, as a research tool, its aim is to generate original information on human diseases and on efficacy and mechanisms of therapies and to help drug development and discovery. In the latter case, the opportunity exists to use in vivo molecular imaging as a translation from laboratory-based studies of molecular, cellular and animal studies to corresponding investigations in diseased tissues in life. If developed, this has considerable implications for providing molecular biologists with insight into the functional sequela of genetic perturbations and molecular biology-based therapies as well as accelerating pre-clinical drug discoveries into the clinic.

This chapter does not attempt to review the field of PET and its applications, but concentrates on the specific challenges that exist to ensure that its role in molecular medicine is fully developed and exploited. In the first instance, these rest on refining the methodology of PET with the goal of optimising the accuracy of generated functional images of a molecular pathway or interaction. These depend on both the quality of the collected data and its subsequent processing and analysis. The logistics of radiolabelling with rapidly decaying radioisotopes such as ^{11}C ($t_{1/2}$, 20.1 min) and the handling of large amounts of kinetic data need to be solved to ensure specificity and the full exploitation of inherent temporal and spatial resolutions. The challenge of justifying this specialised area of molecular imaging, within the clinical domain, needs careful consideration in order to avoid overlap with other methodologies both within in vivo imaging and in vitro assays. It will be seen that a number of hurdles need to be overcome to make impacts in therapeutically intractable diseases such as schizophrenia, Alzheimer disease, heart failure and carcinomas.

Finally, if it is not met, the challenge of bridging the cultural gap between laboratory- and clinically based science may represent the greatest threat to this field reaching its full potential for translation between molecular biology, pre-clinical drug discovery and the patient.

Key aspects of the future strategy are itemised to ensure that the challenges facing PET are responded to.

7.2 The Imaging Science of PET

The imaging science of PET covers the strategies for advancing methodology to improve the accuracy of functional images generated using PET of regional tissue molecular interactions and pathways (Jones 1996b; Myers et al. 1996). Key components are the selection of the tracer molecule, its site of radiolabelling and optimisation of its specificity and selectivity for the molecular process of interest. This is followed by recording scan data in humans with high spatial and temporal resolution for which maximum sensitivity is sought. Processing of recorded data prior to tomographic reconstruction centres on reducing noise due to corrections for random and scattered coincidences. Finally, the temporal data needs to be analysed with kinetic models to derive functional parameters of interest. The accuracy of the final, functional image is dependant on all these stages. Each need to be considered whenever a novel clinical research question is being asked of PET to ensure the methodology is able to provide the answer that is sought.

All the components within this imaging science are currently undergoing refinement. The cyclotron technology for producing short-lived positron-emitting isotopes is well advanced, using compact shelf-shielded machines that are compatible with clinical environments. Most advances focus on improving the efficiency of the target vessels irradiated to produce radioisotopes with high yields and specific activities, i.e. the ratio between the radioisotope and its stable isotopic form. PET scanner technology has advanced to maximise the solid angle for detecting the photons emitted from selected organs such as the brain and heart. The detectors used fully capture the high-energy 511-KeV photons associated with positron emission. Present effort is focused on how to accurately correct for the presence of scattered coincidences which impose a background upon the data set. New detector material has been developed which promises to result in lower losses due to dead time effects at high count rates and random coincidence registration. Data collection strategies, involving rapid access to large computer storage, are being addressed to deal with the high data flow rates present during PET kinetic studies. Data processing is advancing on two fronts. Prior to reconstruction, methods are being refined in order to minimise the statistical deterioration resulting from corrections of scattered and random coincidences. Secondly, sophisticated reconstruction techniques which require the use of parallel computing are being explored to produce more accurate tomographic data.

The kinetic analysis of high temporal resolution data sets is faced with a number of challenges. Traditionally, compartmental model-based analyses have been adopted on a "region of interest" basis. These computations tend to be statistically unstable in that the statistical content at the voxel level in the image is insufficient for a full compartmental fitting involving three or more rate constants. As a result, surrogate parameters are being developed, such as the tracer's volume of distribution and binding potential within the tissue. Furthermore, data-led approaches are being explored in order to find, at the voxel level, the number of statistically identifiable kinetic components that actually reside within the data set (Cunningham and Jones 1993). This is particularly important when dealing with diseased tissues where conventional models, accurate for normal tissues, begin to break down, e.g. in tumours, Alzheimer disease and myocardial infarction. In an attempt to maximise statistical information, population-based kinetics are currently being developed against which the results of a single patient study can be compared to identify regional perturbations in a molecular pathway or interaction. When formulating kinetic analysis, it is important to capitalise on the whole of the kinetic data set and not just on some end-point measure. The goal here is to use all the available statistical information that results from the radiation exposure to the subject under investigation. To compliment these advances in data analysis, statistical techniques are being introduced to interrogate data sets objectively in order to identify focal changes in molecular interactions both from the normal condition and following treatment.

Last but not least, a further methological challenge is the selection of the tracer molecule itself. A major restraint in the field of molecular imaging is the dearth of selective and specific molecules for in vivo studies of molecular interactions and pathways. There are severe demands on such probes. Not only is selectivity necessary, but non-specific binding to other tissue components should be low. Furthermore, the tracer's metabolism needs to be minimised during the time course of the study, since radiolabelled metabolites contaminate the recorded signal, requiring additional strategies for their correction. These last three restraints are often not a problem in traditional in vitro binding studies and hence require additional strategies not developed outside of the in vivo molecular imaging field.

Implementing advances in molecular imaging requires a multi-disciplinary approach encompassing chemistry, biology, pharmacology, physics, mathematics and computing science. These need to be complimented and cross-fertilised by clinical specialists who define research questions and goals. Iteration is required between methodology and would-be clinical research users of this imaging modality (Jones et al. 1996). Innovation is necessary, and insight, problem-solving and team activity is germane to this field. Even with established methodology, addressing well-defined clinical research questions, the logistics for implementing clinical studies involving rapidly decaying, radio isotopes and the collection of large kinetic data sets is itself a formidable undertaking requiring multi-disciplinary activities.

7.3 Link with Molecular/Cell Biology and Drug Discovery and Development

To date, the above implementations of in vivo molecular imaging have been confined to clinical research and diagnostic environments. However, bearing in mind the future potential of such methodology, it is clear that this confinement is to some extent a weakness. In the first instance, it is useful to consider why basic activities in molecular/cell biology and drug discovery and development should align themselves with in vivo molecular imaging for which PET is the most sophisticated example.

From the outset, it is appropriate to consider in vivo molecular imaging as a means to *translate* between basic laboratory science in molecular/cell biology to related studies within disease tissues in humans. To an outsider, it would appear that, within the pharmaceutical industry, such translation is currently lengthy, expensive and often associated with little or modest feedback, based upon the clinical effects of a novel pharmaceutical agent, to scientists within drug discovery teams. Once a compound is a candidate for development in the clinical domain, a series of somewhat ballistic programmes are established, namely phases 1, 2 and 3. The criteria used at the critical decision-making stages of such drug development are comparatively crude, e.g. toxicity and early clinical response data. As a result of this clinical trial phase, some modest feedback to the drug discoverers comes about, but often some years later. However, this is often in the form of soft clinical signs,

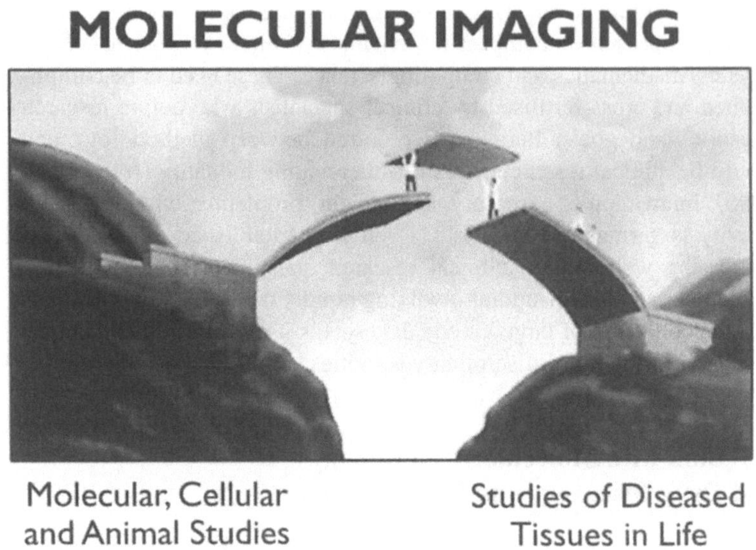

Fig. 1. Molecular imaging. (This is a variation of a figure originally defined for the field of nuclear medicine by H.N. Wagner 1995; see also "Molecular Nuclear Medicine: From Genotype to Phenotype," this volume)

perhaps supplemented with assays of some biological material such as plasma levels of the drug. One understands that within the research and development budgets of large pharmaceutical companies, approximately 80% is spent on the development side, often with little scientific pay-off, while only 20% is spent on discovering new agents. Although many of these development costs are unavoidable (e.g. toxicity studies and population trials of clinical efficacy), it is clear that these need to be minimised in order to mobilise more resources for discovery.

With respect to the field of molecular biology, there does appear to be a large gap between fundamental discoveries and how these translate to cell and whole organ physiology and pathophysiology (Weinberg 1996). From the above case for more direct and effective translation, it can be concluded that the ability to use in vivo molecular imaging to provide pharmacokinetic profiles and functional measurements for pharmacodynamic end-points will provide the data needed to bridge this gap (Fig. 1).

7.4 The Challenge to Effect Translation

The challenge to effect translation should be seen at two levels. The exponents of PET face an ongoing shortage of selective in vivo tracers. Most, if not all candidate molecules reside within the huge libraries of chemical compounds that exist within the pharmaceutical companies. For the PET field to advance and realise its full potential, this resource needs to be accessed and interrogated for potential tracer compounds. To help this, a bioinformation strategy is needed whereby templates that make for a good in vivo tracer are used, together with computer modelling, to help interrogate chemical libraries.

The largest hurdle is the culture gap between drug discoverers, who have their own, well-tried and proven ex vivo and in vitro methods to define a compound's pharmacological activity. Gradually, they are beginning to realise the potential that in vivo tracer studies offer. Drug developers are more advanced in their consideration in that they have seen ways in which clinical trials can be accelerated when using in vivo assays (Comar 1995). However, the fact remains that very few people in the pharmaceutical industry have first-hand experience of PET and hence are not in the position to fully champion its use within the industry. An additional inducement is that animal PET studies provide a link with drug discovery teams, and small-animal PET scanners have been developed specifically for this purpose (Hume et al. 1996). The case is suggested for these being the focus of drug company research within an established PET centre. Other thoughts have already advanced to the stage of undertaking animal PET facilities within a pharmaceutical company itself.

Hence the challenge that presents itself in order to realise the potential of PET for clinical research and drug discovery and development is to forge links with drug discoverers. Not only will this direct the strength of the imaging methodology into an exciting and sophisticated area with huge potential, but it will also provide the much sought after tracer compounds and the pharmacological and medicinal chemistry expertise needed to advance this field for clinical research.

References

Comar D (ed) (1995) PET for drug development and evaluation. Kluwer, Dordrecht, pp 301–326

Cunningham VJ, Jones T (1993) Spectral analysis of dynamic PET studies. J Cereb Blood Flow Metabol 13:15–23

Hume SP, Lammertsma AA, Myers R, Rajeswaran S, Bloomfield PM, Ashworth S, Fricker RA, Torres EM, Watson I, Jones T (1996) The potential of high resolution positron tomography to monitor striatal dopaminergic function in rat models of disease. J Neurosci Methods 67:103–112

Jones T (1996a) The role of position emission tomography within the spectrum of medical imaging. Eur J Nucl Med 23 (2):207–211

Jones T (1996b) the imaging science of positron emission tomography. Eur J Nucl Med 23 (7):807–813

Jones T, Silbersweig DA, Stern E, Schnorr L, Seaward J, Clark JC, Lammertsma AA, Grootoonk S (1996) The development of *in vivo* tracer methods to obtain new information about human disease: a study of the hallucinating brain. Eur J Nucl Med 23:332–335

Myers R, Cunningham V, Bailey D, Jones T (eds) (1996) The quantification of brain function using PET. Academic, San Diego

Wagner HN (1995) A new era of certainty. J Nucl Med 36 (8):13N–28N

Weinberg RA (1996) E2F and cell proliferation: a world turned upside down. Cell 85:457–459

8 Pharmacological Studies with Positron Emission Tomography

J.J. Frost

8.1 Introduction 91
8.2 Targeting of Neurochemical Sites 92
8.3 Selection of Labeled Drugs 93
8.4 Drug Development 95
8.4.1 Preclinical Phase 95
8.4.2 Clinical Approval 96
8.4.3 Post-Approval Applications 96
8.5 Determination of Neurochemical Mechanisms of Disease 97
8.5.1 Brain Opioid Receptors in Cocaine Abusers 97
8.5.2 Dopamine Transporter in Parkinson's Disease 98
8.6 Conclusion 99
References 100

8.1 Introduction

Positron emission tomography (PET) scanning of the brain had its origins in the measurement of regional blood flow and substrate metabolism. While these types of measurements continue to play an active role in the localization of brain function and in the clinical management of patients with diseases such as epilepsy, brain tumor, and dementia, the flow metabolism methods have limitations. The results of blood flow functional activation studies have provided a wealth of information on regional brain activation following sensory and cognitive stimuli, but mechanistic information regarding the functional role of neurotransmitters, neuroreceptors, and other neuromarkers in increasing or decreasing neuronal activity is not provided. Similarly, measurement of regional brain metabolism is useful for identifying and localizing functional abnormalities, differential diagnosis, and following the affects of treatment, but information regarding mechanisms of disease at a neurochemical level is lacking in these types of studies. For example, epilepsy is known to be associated with specific patterns of abnormal regional glucose metabolism, but epilepsy is not a disorder of altered carbohydrate metabolism. Rather, there are a series of inhibitory and excitatory neurotransmitters and their receptors that are thought to play a role in the origin of seizures. In contrast to flow metabolism measurements, which do not provide a link to pharmacotherapeutic interventions, knowledge of neuroreceptor and neurotransmitter abnormalities in human disease can provide insight into appropriate therapies, since many centrally acting drugs act at specific neurochemical sites, such as neuroreceptors. Moreover, while flow metabolism methods give us information about *where* things happen in the brain, more chemically specific tracers for PET would tell us *how* things happen.

8.2 Targeting of Neurochemical Sites

There are a number of neurochemical sites which are being targeted using PET. These include neurotransmitter synthesis, neurotransmitter packaging into synaptic vesicles, neurotransmitter metabolism by monoamine oxidase (MAO), neurotransmitter reuptake, neuroreceptor binding, and, more recently, intracellular second messengers and similar sites such as adenylate cyclase, N proteins, and G proteins. As PET imaging has evolved to more chemically specific tracers, the sensitivity in detecting early biochemical change has improved. For example, in Parkinson's disease there is an 80% or greater loss of dopamine and dopaminergic nerve terminals in the nigrostriatal pathway, yet glucose metabolism is essentially normal in the caudate–putamen and other brain regions. However, with more chemically specific tracers such as ^{18}F-FDOPA and ^{11}C-WIN 35,428 (see below), very early loss of dopaminergic nerve terminals can be detected in Parkinson's disease. Although dopminergic nerve terminals decline markedly in the striatum of such patients, they represent only a small fraction of all nerve terminals in this region; thus the small fractional decline in neuronal activity cannot be detected by PET metabolic methods, in spite of the severe functional abnormalities which these patients experience.

8.3 Selection of Labeled Drugs

The selection of labeled drugs for imaging neuroreceptors by PET is not an easy task, since most pharmacologically active drugs do not have the requisite properties for labeling neuroreceptors in vivo under conditions amenable to PET imaging. For example, in the case of imaging brain opioid receptors, effective analgesics such as morphine and dihydromorphine do not preferentially label opioid receptors in vivo (i.e., in relation to nonspecific binding sites), in large part because the affinity of these drugs for opioid receptors in vivo is relatively low and, accordingly, dissociation from receptors is fast. Imaging of opioid receptors in vivo by PET could only be accomplished with the drug [^{11}C]carfentanil, which has subnanomolar affinity to the opioid receptor. [^{11}C]Carfentanil, once bound to opioid receptors in vivo, dissociates very slowly in comparison to nonspecifically binding drugs. Other factors that are important in choosing ligands for PET imaging of receptors are ease of labeling, metabolic fate, blood–brain barrier transport, and nonspecific binding. The differential brain clearance kinetics results in a preferential accumulation of [^{11}C]carfentanil in opioid receptor-rich brain regions about 30 min after intravenous administration. Under these conditions, [^{11}C]carfentanil binding is high in the thalamus, caudate–putamen, and amygdula, intermediate in cerebral cortex, and very low in the occipital cortex, a region which can be used to measure nonspecific binding (Frost et al. 1985).

Since 1983, a number of neuroreceptors have been localized and quantified by PET, including the opioid dopaminergic benzodiazepine and muscarinic cholinergic, serotonergic, and histimine receptors (Frost 1986). Although the PET ligands for these receptors are often very potent, high-affinity drugs studies can most often be conducted at trace drug levels (approximately 0.1 g/kg or less), levels that result in very low receptor occupancies and, accordingly, few if any pharmacologic effects. This is a big advantage in terms of the safety aspect of these studies, but also means that drugs can be used which would not otherwise be approved or appropriate for human use. As discussedbelow, the use of PET radiopharmaceuticals with highly specific activity under these conditions is an opportunity for drug discovery and development.

Quantification of receptor binding is possible using a variety of methods, many of which rely on multicompartmental modeling. Since receptor binding is a saturable process, at least two labeled drug doses must be used if the equilibrium dissociation constant (K_D) and receptor number (B_{max}) are to be estimated. In a single PET study on highly specific activity, only the product of the association rate constant and the receptor number can be estimated (konB_{max}) (Price et al. 1993; Sadzot et al. 1991). A simplified approach to receptor quantification is based on the computation of radioactivity counts between regions containing receptors and a region representing nonspecific binding. For example, in the case of opioid receptor binding, the occipital cortex can be used as an internal standard for nonspecific binding. This approach is more suitable for ligands that attain near-equilibrium binding conditions during the time course of the PET study. This approach can be validated against a multicompartmental analysis to demonstrate that the simplified ratio method yields accurate estimates of regional binding (Frost et al. 1989).

Saturation of neuroreceptor binding in vivo can be demonstrated using the unlabeled form of the radioligand or, alternatively, by the use of other unlabeled drugs. For example, [^{11}C]carfentanil binding can be inhibited by the administration of intravenous nalaxone or oral naltrexone, two opiate antagonists. Receptor binding inhibition or displacement studies can be performed (Frost and Wager 1990). In the inhibition paradigm, unlabeled drug is given prior to the administration of the radiopharmaceutical. In the displacement study, radioligand is administered and allowed to bind to receptors, and an unlabeled drug is administered to induce displacement from receptors. The displacement paradigm can be used when one needs to know whether the binding is reversible or not. For example, some receptors are internalized following binding, in which case the labeled drug might not be freely displaceable.

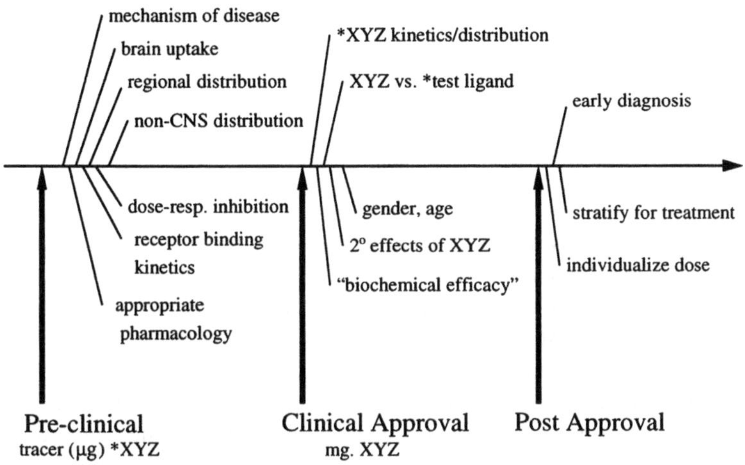

Fig. 1. Potential uses of PET in drug development are summarized for three phases: pre-clinical development, during the clinical approval process, and after approval. The figure shows the development process for the hypothetical drug *XYZ* which can be radiolabeled for PET studies (**XYZ*)

8.4 Drug Development

One established use of PET imaging of drug-binding sites is in the development and use of new pharmacotherapeutic agents. The possible uses of PET in drug development are depicted in Fig. 1. The drug development process is divided into three phases: the preclinical phase, the phase of clinical approval, and the post-approval phase.

8.4.1 Preclinical Phase

During the preclinical phase, milligram mass doses of a drug are not yet approved for human use, but one can generate information on the in vivo distribution of the new drug by radiolabeling it for PET scanning. PET scanning can then be used not only to evaluate the in vivo kinetics and receptor binding profile of a new drug, but also to determine whether there are abnormalities in the regional distribution of the receptor in various diseases. Documentation of regional receptor abnormalities in disease can represent an important component of the rationale for further clinical development of a drug.

8.4.2 Clinical Approval

Following the approval to administer milligram or therapeutic quantities of the drug to humans, another set of PET experiments might be conducted. One might wish to determine whether there are any substantial differences in the distribution of the drug when comparing microgram and milligram quantities. Unlabeled drug can also be tested against well-validated selective PET ligands for various neuroreceptors if there is a question about the subtype or receptor selectivity of the new drug. The secondary effects of the drug on the regulation of various neuroreceptors can also be evaluated as a way to gain insight into mechanism of action. Since it is well established that drug effects and side effects can be dependent on gender and age, one might want to determine at the stage of clinical approval whether there are significant gender- or age-associated differences in drug–receptor binding. The results of such studies would be helpful in designing future clinical trials to determine drug action as a function of gender or age or, conversely, to provide evidence that no such studies are needed. Finally, in diseases in which a well-defined neuromarker of progression is known, PET scanning could be used to determine whether a drug (e.g., a neuroprotective drug) is effective in slowing disease progression. This could be a viable alternative to clinical trials for determining drug efficacy with the potential advantages of lower costs and more rapid generation of results.

8.4.3 Post-Approval Applications

Following the approval of a drug for clinical use, PET scanning could still have a role in refining the use of the drug. One alternative is to extend the use of the radiolabeled drug to a diagnostic imaging agent which could be used for early diagnosis and differential diagnosis of specific disorders in which the drug might be employed as a pharmacotherapeutic agent. Insofar as the disorders in question might represent a biochemically heterogeneous group, imaging could be used for treatment stratification (i.e., only those patients with a specific receptor abnormality would be treated with the drug). Conceivably, PET scanning could be used to individualize drug dose in order to maximize the therapeutic response and minimize serious side effects. For example, PET studies using dopaminergic antipsychotics have shown that a receptor occupancy of about 70% is needed for a therapeutic response and that beyond 80% occupancy there is a high likelihood of tardive dyskinesia (Farde et al. 1992).

8.5 Determination of Neurochemical Mechanisms of Disease

There are a number of examples of using PET to determine neurochemical mechanisms of disease, but a thorough review is beyond the scope of this article. In the following, two examples will be briefly discussed.

8.5.1 Brain Opioid Receptors in Cocaine Abusers

One recent illustrative example is the investigation of regional abnormalities of brain opioid receptors in cocaine abusers (Zubieta et al. 1996). The primary mechanism of cocaine is to inhibit dopamine reuptake by the dopamine transporter and to elevate intrasynaptic dopamine concentration. However, this elevation of dopamine causes indirect affects on the endogenous opioid system by acutely increasing release of endogenous opioids such as the enkephalins. Activation of the endogenous opioid system is a powerful mechanism for reinforcement and, accordingly, some of the reinforcing effects of cocaine might be mediated through the endogenous opioid system, as is the case for drugs such as heroin. A study was conducted using PET to examine regional abnormalities in opioid receptors in chronic cocaine abusers. The results demonstrated that there were regional elevations in m-opioid receptor binding in regions associated with reinforcement, such as the frontal cortex, the cingulate cortex, and the amygdala. Interestingly, although in animal models acute cocaine elevates levels of endogenous opioids, in chronic cocaine administration there is a reduction in endogenous opioid release during periods between acute administrations. Accordingly, the elevations in regional opioid binding may represent an upregulation response of opioid receptors due to chronic reductions in enkephalin release.

A second and important finding of this study was that the elevation in opioid receptors in many regions was related to the amount of craving for cocaine that these subjects experienced at the time of the PET scan. This finding demonstrated that upregulation of opioid receptors was higher in subjects who experienced more craving for cocaine. This finding can be explained in terms of an imbalance between upregulated opioid receptors and reduced release of enkephalin during periods between self-administration of cocaine. This imbalance may contribute to the phenomenon of craving, which is relieved when subjects administer cocaine and a subsequent release of endogenous opioids occurs. The findings of this study demonstrate that biochemical mechanisms of disease can be elucidated using PET, which, in the case of cocaine abuse, may lead to new pharmacotherapeutic approaches that target the endogenous opioid system. Furthermore, this study shows the value of combining PET biochemical studies with behavioral measures of disease. The observation of a close relationship between a biochemical abnormality and a behavioral phenomenon strengthens the case that the biochemical finding represents an important mechanism of the disease rather than an epiphenomenon. Under these conditions, the biochemical parameter could serve as an important surrogate marker for the behavior or even a primary marker of the disease itself.

8.5.2 Dopamine Transporter in Parkinson's Disease

Another example of the use of PET in elucidating mechanisms of disease is the imaging and quantification of the dopamine transporter in Parkinson's disease. The dopamine transporter or reuptake site is an important presynaptic dopaminergic marker which can be labeled and imaged using PET and the potent uptake inhibitor ^{11}C-WIN 35,428 (Frost et al. 1993). The initial studies show that, in a relatively mild form of Parkinson's disease (Hoehn and Yahr stage II), there were marked declines in dopamine transporter binding in the caudate and putamen of patients with Parkinson's disease. The next phase of the research was to measure reductions in dopamine transporter binding and stage I patients with hemi-Parkinson's disease. In these patients, one side of the body is asymptomatic, thus presenting an opportunity to examine the biochemical abnormalities in the basal ganglia corresponding to the asymptomatic side of the body (Ilgin et al. 1995). In these studies, not only was there a reduction in the caudate–putamen dopamine transporter binding corresponding to the symptomatic side of the body, but there was also a reduction of 30%–50% in the caudate–putamen corresponding to the asymptomatic side of the body. These results demonstrate that PET methods are sensitive and appropriate for detecting presymptomatic biochemical abnormalities in disease. They suggest that PET imaging of the dopamine transporter could be used as a diagnostic imaging method for detecting Parkinson's disease at an early stage, thereby permitting early therapy with neuroprotective agents, such as MOA inhibitors. From the standpoint of drug development, these data also open up the possibility of conducting clinical trials of new neuroprotective agents using PET where the end point is the reduction in the loss of dopamine transporter binding rather than a change in a clinical point such as tremor or bradykinesia. Using PET as a method to determine "biochemical efficacy" of neuroprotective drugs, one might be able to conduct a clinical trial more quickly, with fewer subjects, and less expensively compared to a traditional clinical trial with symptom end points.

8.6 Conclusion

In summary, PET imaging of regional brain pharmacology can play an important role in understanding neurochemical mechanisms of disease and can therefore have an important impact on the selection and testing of new drugs. Disease mechanisms in human CNS disorders have historically played a relatively small role in drug development since, prior to PET imaging of neuroreceptors and other markers, it was not possible to measure regional human neurochemistry in vivo. Nor has it been possible, until recently, to study new CNS drugs at their sites of action (e.g., receptors, enzymes, transporters). PET now offers the possibility of designing and selecting new drugs based on neurochemical principles rather than behavior and testing of the drugs by measuring their kinetics and distribution in the brain as opposed to peripheral organs. Expanded applications of PET to drug development will likely play an important role in shortening the drug development process, reducing costs, and improving drug efficacy.

Acknowledgments. The author wishes to thank Ms. Alexis J. Simich for her technical support. This work was supported by NIDA grant # 1RO1-DA09479, NIA grant # 3R01-AG13194, and NIMH grant # 5R01-MH51141.

References

Farde L, Nordstrom AL, Wiesel FA, Pauli S, Halldin C, Sedvall G (1992) Positron emission tomographic analysis of central D1 and D2 dopamine receptor occupancy in patients treated with classical neuroleptics and clozapine. Relation to extrapyramidal side effects. Arch Gen Psychiatr 49 (7):538–544

Frost JJ (1986) Studies of neurotransmitters and neuroreceptors by emission computed tomography. Trends Pharmacol Sci 7:490–496

Frost JJ, Wager HN Jr (eds) (1990) Quantitative imaging: neuroreceptors, neurotransmitters, and enzymes. Raven, New York

Frost JJ, Wagner HN Jr, Dannals RF, Ravert HT, Wilson AA, Burns HD, Wong DF, McPherson RW, Rosenbaum AE, Kuhar MJ, Snyder SH (1985) Imaging opiate receptors in the human brain by positron tomography. J Comput Assist Tomogr 9:231–236

Frost JJ, Douglass KH, Mayberg H, Dannals RF, Links JM, Wagner HN Jr (1989) Multicompartmental analysis of [11C]-carfentanil binding to opiate receptors in humans measured by positron emission tomography. J Cereb Blood Flow Metab 9:398–409

Frost JJ, Rosier AJ, Reich SG, Smith JS, Ehlers MD, Snyder SH, Ravert HT, Dannals RF (1993) PET imaging of the dopamine transporter with 11C-WIN 35,428 reveals marked declines in mild Parkinson's Disease. Ann Neurol 34:423–431

Ilgin N, Zubieta JK, Reich SG, Ravert HT, Dannals RF, Frost JJ (1995) Preclinical diagnosis of Parkinson's disease with C11-WIN 35428 PET imaging of the dopamine transporter. J Nucl Med 36:98P

Price JC, Mayberg HS, Dannals RF, Wilson A, Sadzot B, Rattner Z, Kimball A, Feldman M, Frost JJ (1993) Measurement of benzodiazepine receptor number and affinity using [11C]-flumazenil and positron emission tomography: comparison of kinetic and equilibrium analyses. J Cereb Blood Flow Metab 13:656–667

Sadzot B, Price JC, Mayberg HS, Douglass KH, Dannals RF, Lever JR, Ravert HT, Wilson AA, Wagner HN Jr, Feldman MA, Frost JJ (1991) Quantification of human opiate receptor concentration and affinity using high and low specific activity 11C-diprenorphine and positron emission tomography. J Cereb Blood Flow Metab 11:204–219

Zubieta JK, Gorelick DA, Ravert HT, Dannals RF, Frost JJ (1996) Increased mu opioid receptors binding detected by PET in cocaine dependent men is associated with cocaine craving. Nat Med 2 (11):1225–1229

9 Use of Complex Carbohydrates as Targets and Ligands for Imaging

R. Tauber, S. Emig, B. Volz, C. Fieger, and C. Bührer

9.1	Introduction	104
9.2	Structures, Classification, and Distribution of Complex Carbohydrates	104
9.3	Biosynthesis and Postbiosynthetic Modifications	111
9.3.1	Biosynthesis of *N*-Linked Oligosaccharides	112
9.3.2	Biosynthesis of *O*-Linked Oligosaccharides	114
9.3.3	Factors that Influence Oligosaccharide Biosynthesis	114
9.4	Biological Functions of Glycoprotein Glycans	116
9.4.1	Role in Protein Folding, Conformational Stability, and Solubility	116
9.4.2	Barrier and Structural Functions	117
9.4.3	"On–Off" and "Tuning" Functions	117
9.4.4	Oligosaccharide–Receptor Interactions	119
9.5	Carbohydrate-Mediated Cell Adhesion	119
9.5.1	Selectin–Carbohydrate Interactions	120
9.5.2	Carbohydrate Selectin Ligands	120
9.5.3	Glycoprotein Selectin Ligands	121
9.5.4	Therapeutic Inhibition of Carbohydrate–Selectin Interaction	122
9.6	Changes of Glycosylation in Disease	124
9.7	Carbohydrates as Ligands for Targeting and Imaging	126
References		127

9.1 Introduction

Complex carbohydrates were long considered to be biological compounds that lack any biological specificity, serving merely as structural components of cells, of the extracellular matrix, and of connective tissues, as mucins on epithelial surfaces, or as a source of energy. Only within the last three decades has it become evident that complex carbohydrates form a highly diverse and complex class of biomolecules that are almost ubiquitous in any living organism and that mediate a large number of biological functions, ranging from purely structural roles to such complex processes as cell–cell interactions operative in the immune system or during embryonic development (for selected monographs and reviews see Rademacher et al. 1988; Kobata 1992; Varki 1993; Lis and Sharon 1993; Fukuda and Hindsgaul 1994; Montreuil et al. 1995). Moreover, inherited and acquired carbohydrate disorders have been discovered to play key roles in the etiology and progression of various types of disease, including metabolic disorders, immunological diseases, infections, and malignancy (selected references: Kornfeld 1986; von Figura and Hasilik 1986; Hakomori 1989; Etzioni et al. 1992; Furukawa and Kobata 1991; Muramatsu 1993; Ofek and Sharon 1990; Jaeken et al. 1993). Apart from contributing to the understanding of basic pathological mechanisms, studies on glycosylation changes in disease have led to the discovery of distinct carbohydrate structures as diagnostic markers of disease processes and to the development of novel diagnostic procedures and of new diagnostic and therapeutic agents (Young et al. 1991; Turner 1992; Muramatsu 1993; Mulligan et al. 1993; Parekh and Edge 1994). It is the aim of this chapter to briefly review the structures and functions of complex carbohydrates, of glycosylation changes in disease, and to describe the use of carbohydrate epitopes as targets and ligands for diagnostic and therapeutic agents.

9.2 Structures, Classification, and Distribution of Complex Carbohydrates

Complex carbohydrates are characterized by a high degree of structural diversity, which is one of the molecular bases for their biological functions (Fig. 1). The first source of structural diversity derives from the

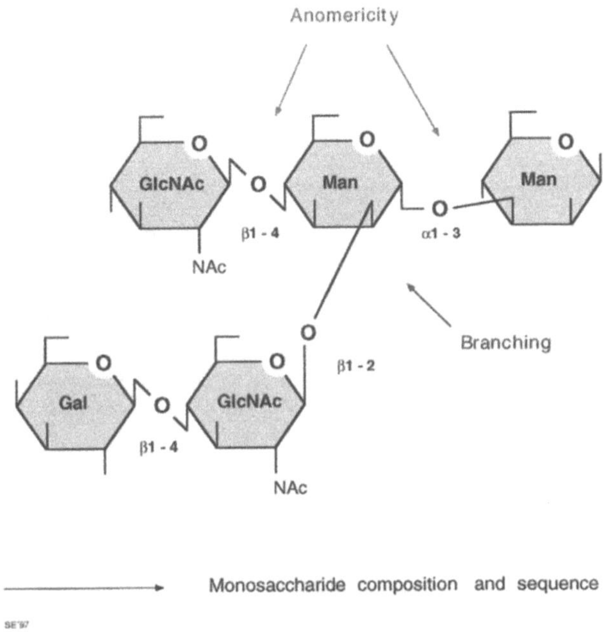

Fig. 1. Determinants of structural variability of complex carbohydrates

chemical nature of the oligosaccharide monomers in that monosaccharides can be arranged in various compositions and sequences within an oligo- or polysaccharide. For a long time, oligo- and polysaccharides were thought to be constituted by D-galactose, D-mannose, D-glucose, L-fucose, D-xylose, L-arabinofuranose; the amino sugars *N*-acetyl-D-glucosamine and *N*-acetyl-D-galactosamine; and sialic acids. Novel analytical methods have led to the discovery of additional monosaccharides which are components of complex carbohydrates (for review, see Vliegenthart and Montreuil 1995). As an example, according to present knowledge, the group of sialic acids that has pivotal biological functions comprises 35 different members (Schauer 1991). Secondly, as monosaccharides usually have three or four free hydroxyl groups, the oligosaccharide monomers may be linked to each other at different positions. In addition to different linkages between sugars arranged in a linear se-

Table 1. Types of glycoconjugates in eukaryotic cells

Type	Glycans	Distribution
Glycoproteins	*N*-and *O*-linked mono-, di- and oligosaccharides	
Plasma/serum glycoproteins		Plasma/serum, extracellular fluids, effusions
Membrane glycoproteins		Cell surface membrane and intracellular membranes
Matrix glycoproteins		Connective tissues, extracellular matrix
Glycoprotein hormones		Plasma/serum
Cytoplasmic/nucleoplasmic glycoproteins		Cytoplasm, nucleus
Proteoglycans	Glycosaminoglycans	Connective tissue, extracellular matrix, cell surface membrane
Glycolipids	Mono-, di-, oligosaccharides	
Cerebrosides		Cell membranes
Sulfatides		
Gangliosides		

quence, this raises the possibility of branching (Fig. 1). Thirdly, since *O*-glycosidic linkages may be either in α or β position, each disaccharide can exist in two anomeric conformations. The structural diversity is even extended by a significant but limited motional flexibility of oligosaccharide conformation (Homans 1995). As a result the various monosaccharides in complex carbohydrates, at least theoretically, may exhibit a much greater structural variability than proteins or nucleic acids and may encode biological information.

Although an astronomical number of carbohydrate structures are theoretically conceivable, certain restrictions have been observed which form the basis for several levels of classification. Complex carbohydrates are normally linked covalently to proteins and lipids, constituting the family of glycoconjugates (Table 1). In *glycoproteins* mono-, di-, oligo-, or polysaccharides are covalently bound to a protein. Individual glycoproteins often carry several carbohydrate units which can be attached at different positions of the polypeptide by either *N*-linkage or *O*-linkage (Vliegenthart and Montreuil 1995). *Proteoglycans* are a subclass of glycoproteins distinguished by the nature, quantity, and arrangement of their sugar side chains, which are termed glycosaminoglycans. Glycosaminoglycans are much longer (100–200 monosaccharide residues) than glycans in other glycoproteins and form long, unbranched chains composed of characteristic disaccharide repeats. Glycosaminoglycans contain anionic residues such as sulfate and uronic acid, giving them a highly negative overall charge. With respect to the basic disaccharide units four different types of glycosaminoglycans can be distinguished: (1) hyaluronan, (2) chondroitin and dermatan sulfate, (3) heparan sulfate and heparin, and (4) keratan sulfate. Except for hyaluronan are all found attached to proteins. Although composed of repetitive motifs, glycosaminoglycans of a given proteoglycan exhibit marked heterogeneity within as well as between the individual oligosaccharide chains. This is caused by sulfate substitution and epimerization (Kjellen and Lindahl 1991).

The *glycolipids* constitute a second class of glycoconjugates. Glycolipids are constituents of the outer surface of the plasma membrane, but are also present in distinct intracellular membrane compartments. In glycolipids of animal cells one glycan composed of D-galactose, D-glucose, *N*-acetyl-D-galactosamine and *N*-acetylneuraminic acid is linked to ceramide. Simple glycolipids with only one neutral sugar residue forming their headgroup are called cerebrosides. The most complex of the glycolipids are gangliosides, with their glycans containing one or more negatively charged sialic acid residues (for review, see Hannun and Bell 1989; Hakomori 1990).

Glycosylphosphatidylinositol (GPI) membrane anchors represent a third, smaller class of glycoconjugates. They are complex glycophospholipids covalently bound to the C-terminus of a variety of externally oriented plasma membrane proteins (Cross 1990; Ferguson and Williams 1988). The GPI moiety is partly embedded in the outer surface of the plasma membrane and serves to anchor various proteins to the external surface of eukaryotic cells. GPI-anchored proteins include cell-specific antigens, parasite coat proteins, enzymes, cell adhesion molecules, receptors, and transporters. Although, apart from the anchoring function, the exact biological role is not yet known, it is assumed that the GPI anchor may be related to transmembrane signaling, sorting to the apical surface of polarized epithelial cells, and clathrin-independent endocytosis via caveolae. GPI anchors are composed of a lipid portion and a tetrasaccharide which is linked glycosidically to the 6-hydroxyl group of phosphatidylinositol. The terminal sugar is linked via a phosphodiester bond to ethanolamine, which in turn is linked to the α-carboxyl group of the C-terminal amino acid of the protein. GPI anchors exhibit species- and tissue-specific variations in the lipid portion and in sugar substituents joined to the tetrasaccharide core. These additional sugars give rise to a structural diversity in the carbohydrate moiety of GPI anchors.

This chapter will focus on complex carbohydrates attached to the peptide chains of "classical" glycoproteins by either *N*-glycosyl linkages to asparagine or by *O*-glycosyl linkages to serine, threonine, hydroxylysine, or hydroxyproline. "Classical" *N*-linked oligosaccharides have an *N*-acetylglucosaminyl-asparagine bond, with the *N*-acetylglucosaminyl residue being the innermost sugar of a common pentasaccharide core unit Manα1–6(Manα1–3)Manβ1–4GlcNAcβ1–4GlcNAc (Fig. 2). Different mono- and oligosaccharides may be attached to this core. According to the type of the branches linked to the trimannosyl core, three different types of *N*-linked glycans are distinguished (for references, see Kobata 1992; Vliegenthart and Montreuil 1995). Oligomannosyl or high mannose type glycans only have D-mannose residues linked α1–3, α1–6, or α1–2 to the trimannosyl core. By contrast, complex-type glycans have outer antennae composed of *N*-acetyl-D-glucosamine, D-galactose, L-fucose, and sialic acid. Due to branching at the two α-mannosyl residues of the trimannosyl core, complex-type glycans may have up to five outer branches or antennae (Fig. 2). This basic structural diversity of complex *N*-glycans is extended by (a) repeating disaccharide *N*-acetyllactosamine units Galβ1–4GlcNAcβ1–3 in the outer antennae, giving rise to so-called poly-*N*-acetyl lactosamine structures, (b) branching of the *N*-acetyllactosamine repeats to form Galβ1–4GlcNAcβ1–6 (Galβ1–4GlcNAcβ1–3)Gal structures, and (c) numerous variations of the outer antennae, including sulfation and presence of *N*-acetyl-D-galactosamine. A third type of *N*-glycans, termed the hybrid type, has outer branches of the complex type linked to the Manα1–3 arm and D-mannose residues linked to the Manα1–6 arm of

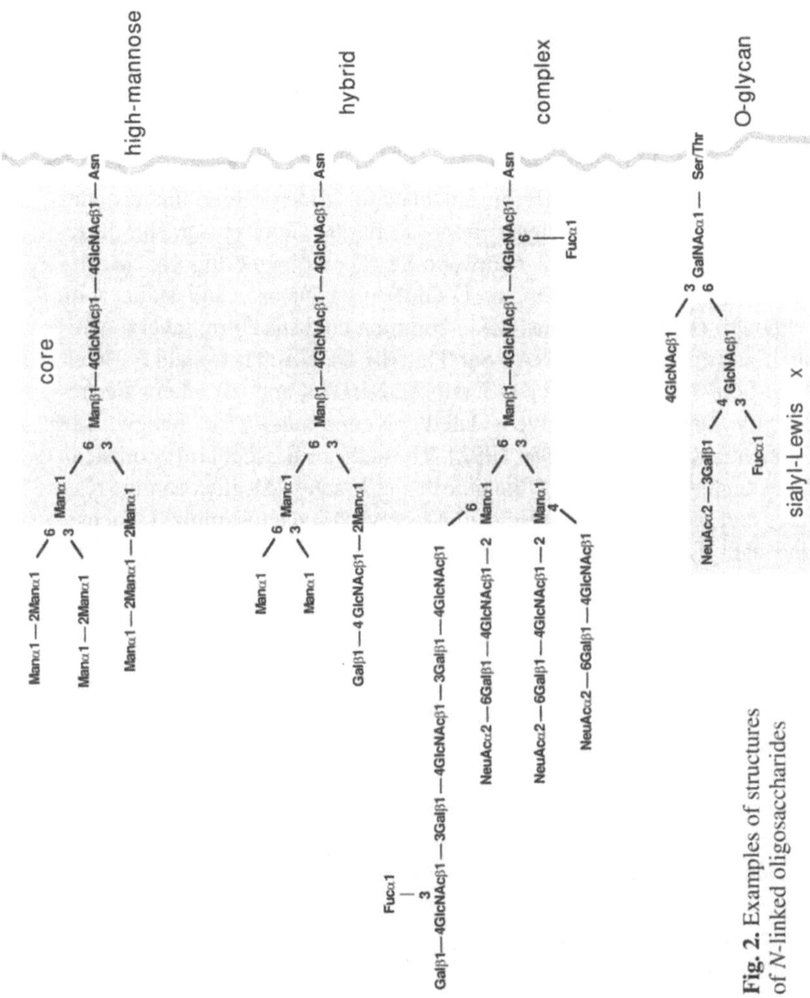

Fig. 2. Examples of structures of *N*-linked oligosaccharides

the trimannosyl core. Frequently, two additional substitutions (d) L-fucose α1-6-linked to the innermost *N*-acetyl-D-glucosaminyl residue of the common pentasaccharide and (e) a "bisecting" *N*-acetyl-D-glucosaminyl residue β1-4-linked to the β-mannosyl-residue of the trimannosyl core are found in both the complex- and hybrid-type *N*-glycans (Fig. 2).

"Classical" *O*-linked glycans attached to serine or threonine residues can be constituted by a single monosaccharide GalNAc-Ser/Thr, disaccharides such as NeuAcα2–6GalNAc-Ser/Thr, or larger oligosaccharides containing D-GalNAc, D-Gal, D-GlcNAc, L-fucose, and sialic acid. Although *O*-linked glycans lack a common core structure, several cores such as (a) Galβ1–3GalNAc-Ser/Thr, (b) GlcNAcβ1–6(Galβ1–3)GalNAc-Ser/Thr, (c) GlcNAcβ1–3GalNAc-Ser/Thr, and (d) others are frequently found on *O*-glycosylated glycoproteins (for review, see Schachter and Brockhausen 1992). These di- or trisaccharide cores can be extended by attaching D-galactose and *N*-acetyl-D-glucosamine residues and by addiing D-galactose, *N*-acetyl-D-galactosamine, L-fucose, and sialic acid as terminal sugars. Apart from these "classical" structural groups of *N*- and *O*-linked glycans, several novel carbohydrate structures have been identified on glycoproteins, including (a) glycosaminoglycans *N*-linked to asparagine on cell surface glycoproteins (Roux et al. 1988), (b) 4-(SO_4) *N*-acetyl-D-galactosamine terminal on *N*-linked oligosaccharides of glycoprotein hormones (Green et al. 1986), (c) *O*-linked-L-fucose and D-glucose on glycoproteins having epidermal growth factor (EGF) homology domains such as tissue plasminogen activator (Harris and Spellman 1993), (d) *O*-linked D-glucose with two additional D-xylose residues (Nishimura et al. 1989), (e) *N*-acetyl-D-galactosamine linked via a phophodiester linkage to complex-type *N*-glycans (Hayes and Varki 1993), and (f) *N*-acetyl-D-glucosamine *O*-linked to serine or threonine of cytoplasmic or nucleoplasmic proteins (Hart et al. 1989).

The biological importance of complex carbohydrates is indicated by their widespread distribution in the organism and by the prevalence or even exclusive presence of distinct types of complex carbohydrates at certain histological and subcellular sites. Extracellular glycoproteins are present in the different tissues and compartments of the organism, including (a) the majority of plasma/serum proteins glycosylated with *N*-linked oligosaccharides, (b) glycoproteins and proteoglycans of connective tissues and of the extracellular matrix in all parenchymal tissues, and (c) *O*-linked sialylated and sulfated oligosaccharides on mucus glycoproteins on the epithelial surfaces of the respiratory, urogenital, and gastrointestinal tracts. At the cellular level, glycoproteins and glycolipids are the predominant components of the cell surface, constituting the glycocalyx that faces the extracellular microenvironment and is in contact with neighboring cells. In the cell's interior the majority of membrane proteins facing the lumen of the endoplasmic reticulum, the Golgi apparatus, and the lysosomes are glycosylated. Very recently, carbohydrate residues on proteins have also been identified in subcellular compartments so far regarded as being devoid of glycoprotein glycans, including *O*-linked *N*-acetyl-D-glucosamine on proteins in the cytoplasm, the nucleoplasm, and the nuclear pore complex (Hart et al. 1989), dermatan sulfates in the nucleus (Hiscock et al. 1994), and, though a matter of controversy glycoproteins in mitochondria (Gasnier et al. 1992).

9.3 Biosynthesis and Postbiosynthetic Modifications

Principally, complex carbohydrates differ from nucleic acids and proteins in that they are not synthesized on a template. Whereas the sequence of the monomeric constituents of DNA, RNA, and proteins is encoded by the base sequence, complex carbohydrates are secondary gene products formed by the sequential and coordinated action of numerous enzymes and transport systems. These include glycosyltransferases that transfer monosaccharides or oligosaccharides from nucleotide sugar donors, dolichylphosphate sugar donors, or the oligosaccharide donor dolichylpyrophosphate-$Glc_3Man_9GlcNAc_2$ to either the phosphorylated polyprenol dolichylphosphate, glycosylation sites on proteins, or acceptor sites on lipid- or protein-bound glycans. Moreover, an array of glycosidases may remove sugar residues from oligosaccharide intermediates during the biosynthetic process. As has been described in several comprehensive reviews (Kornfeld and Kornfeld 1985; Schachter 1986, 1995; Hemming 1995; Verbert 1995; Brockhausen 1995), this coordinated action of glycosyltransferases and processing glycosidases creates the basis for the formation of the wide variety of *N*-linked and *O*-linked glycans. Biosynthesis of both types of glycans takes place in the endoplasmic reticulum, the Golgi apparatus, and the *trans*-Golgi network during transport of newly synthesized glycoproteins to the cell surface.

9.3.1 Biosynthesis of *N*-Linked Oligosaccharides

The biosynthesis of the *N*-linked oligosaccharides of glycoproteins proceeds in severals steps (Fig. 3). Briefly, first a lipid-linked precursor oligosaccharide $Glc_3Man_9GlcNAc_2$ is assembled by stepwise transfer of *N*-acetyl-D-glucosamine, D-mannose, and D-glucose residues to the phosphorylated polyprenol dolichylphosphate in the endoplasmic reticulum. After completing the assembly of the dolicholpyrophosphate bound $Glc_3Man_9GlcNAc_2$, the precursor oligosaccharide is co-translationally transferred by oligosaccharyltransferase to the asparagine residues of suitable sequons Asn-X-Ser/Thr (X can be any amino acid except proline) on the nascent polypeptide chains. Once transferred to the polypeptide chain, the precursor oligosaccharide is subject to stepwise trimming reactions that include removal of glucose and mannose residues by glucosidases I and II and several mannosidases in the endoplasmic reticulum and the *cis*-Golgi to generate $Man_5GlcNAc_2$. At this level of glycoprotein biosynthesis the two inner glucose residues seem to be of particular importance for promoting folding, oligomeric assembly, and quality control of the newly synthesized glycoproteins (Helenius et al. 1997) (see below). Whereas these early steps of oligosaccharide processing are almost identical for all newly synthesized glycoproteins, subsequent steps localized in the Golgi give rise to the numerous individual glycan structures found on mature glycoproteins. The $Man_5GlcNAc_2$ intermediate is the preferred substrate for *N*-acetylglucosaminyltransferase I of the medial Golgi compartment generating $GlcNAcMan_5GlcNAc_2$. This structure can be acted upon by various additional enzymes, including *N*-acetylglucosaminyltransferase III and *N*-acetylglucosaminyltransferase IV, or can move from the medial to the *trans*-Golgi to form different types of hybrid structures. Alternatively, two additional mannose residues can be removed from $GlcNAcMan_5GlcNAc_2$ by mannosidase II in the medial Golgi to form $GlcNAcMan_3GlcNAc_2$, which can be acted upon by *N*-acetylglucosaminyltransferase II, III, or IV or by α6-fucosyltransferase, or can

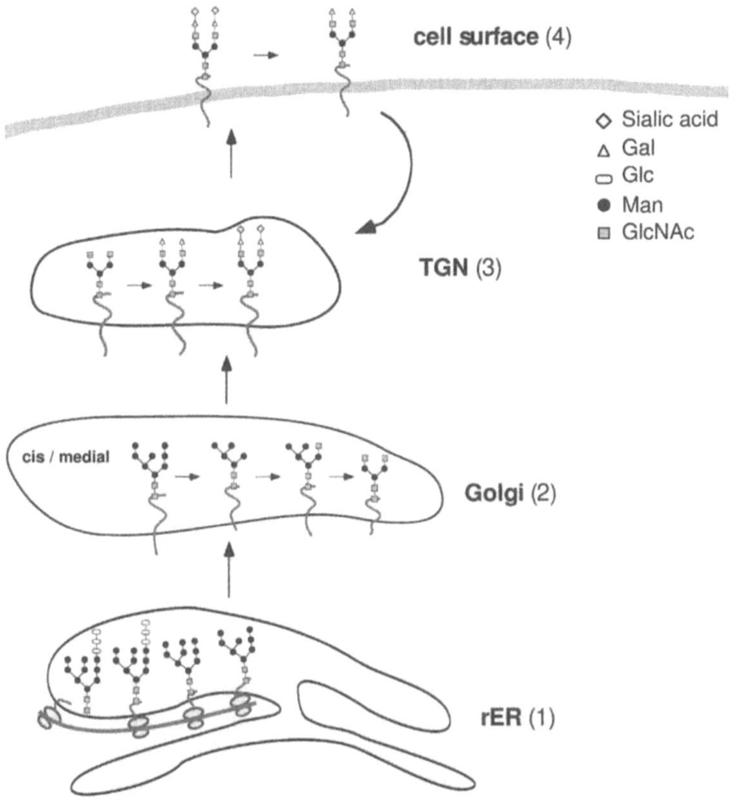

Fig. 3. Biosynthesis of *N*-linked oligosaccharides. *1*, Endoplasmic reticulum (*rER*): cotranslational transfer of $Glc_3Man_9GlcNAc_2$ from DolPP to Asn-X-Ser/Thr acceptor sites on the nascent polypeptide chain by oligosaccharide transferase. Removal of the three glucose and of mannose residues by processing glucosidases and mannosidases. *2*, Golgi apparatus: removal of additional mannose residues by mannosidases I and II, transfer of *N*-acetyl-D-glucosamine residues by GlcNac-transferases I, II, III and IV. Transfer of galactose and fucose residues by galactosyl- and fucosyltransferases. 3, *Trans*-Golgi network (*TGN*): transfer of sialic acid residues by sialytransferases. *4*, Cell surface: stepwise removal of sialic acid, fucose, galactose and mannose residues by exoglycosidases

enter the *trans*-Golgi where the outer chains are extended by galactosyl-, fucosyl-, and sialyltransferases. Since the multiple processing enzymes may act upon the oligosaccharide substrates in various combinations, this nontemplate assembly line is perfectly suited to generate the wide variety of oligosaccharide structures found in secretory and membrane glycoproteins.

9.3.2 Biosynthesis of *O*-Linked Oligosaccharides

The biosynthesis of *O*-linked glycans differs from that of *N*-linked oligosaccharides in that no common lipid-linked oligosaccharide precursor is formed. Instead, the innermost D-GalNAc residue is directly transferred from UDP-GalNAc to Ser/Thr residues on the polypeptide, followed by elongation of the glycan by stepwise transfer of additional sugars. Moreover, *O*-linked glycans are not processed by glycosidases during the biosynthetic process. Similar to the assembly of the outer antennae of *N*-linked glycans, assembly of the different types of *O*-linked glycans occurs along various pathways.

Glycosylation of cell surface glycoproteins, in addition, may be modulated postbiosynthetically, in that sugar residues can be removed from the termini of glycans as well as being re-transferred to preformed glycoproteins on the cell surface or in the Golgi apparatus. These *postbiosynthetic processing reactions* are presumed to present a pathway by which cell surface carbohydrates on preformed glycoproteins are modified in response to endogenous or exogenous stimuli without the necessity for breakdown and de novo synthesis of glycoproteins (Tauber et al. 1994).

9.3.3 Factors that Influence Oligosaccharide Biosynthesis

Despite a certain structural microheterogeneity, individual glycosylation sites on a glycoprotein tend to have oligosaccharides exhibiting characteristic structural features. Hence, mechanisms must be operative that control the formation of distinct oligosaccharide structures at a given glycosylation site of a glycoprotein (Table 2). Cellular factors that guide formation of oligosaccharide structures include (a) the location of

Table 2. Factors that control oligosaccharide biosynthesis

Cellular factors	Polypeptide factors
Expression of glycosyltransferases and glycosidases	Protein conformation
Substrate levels	Quarternary structure
Newly formed glycans	
Nucleotide sugars	Membrane anchoring
Routes and kinetics of intra- and intercompartimental transport	Position of glycosylation sites, steric accessibility of glycans

the various glycosidases and glycosyltransferases in the subcompartments of the endoplasmic reticulum and the Golgi apparatus constituting a glycan "assembly line" (Roth 1995), (b) the activity and substrate specificity of the glycosyltransferases, (c) the concentration of nucleotide sugar donors in the lumen of the Golgi apparatus regulated by specific antiport-transport systems (Abeijon and Hirschberg 1992), and (d) the routes and the velocity of intracellular transport of the biosynthetic intermediates along the assembly line. The level of expression of enzymes and antiport–transport systems is mainly under genetic control at the transcriptional and translational level (Glick 1995). This explains why protein glycosylation is markedly influenced by the genetic program of cells as is reflected by differences in glycosylation between species, individuals, and cell types, as well as by changes in glycosylation during development and differentiation and in response to genetic defects (for review, see Kobata 1992). In addition, the biosynthetic machinery may also be influenced by the cellular and extracellular "microenvironment," e.g., by conditions affected by nutrition or by metabolic diseases, although little is known about these nongenetic influences. Apart from cellular factors, oligosaccharide processing is also influenced by the polypeptide backbone of newly synthesized glycoproteins, i.e., by the conformation and the quarternary structure, which have an impact on the steric accessibility of the glycans for processing enzymes (Hubbard 1988).

9.4 Biological Functions of Glycoprotein Glycans

The oligosaccharides of glycoproteins may have numerous biological functions including rather general roles that do not depend on distinct carbohydrate structures such as their role in protein folding, conformational stability and solubility, more specific roles related to distinct biophysical features of the glycans such as barrier and structural functions, and highly specific interactions with receptors having binding specificity for refined shapes formed by the oligosaccharide ligand. As outlined in a recent comprehensive review (Varki 1993), it is difficult to predict specific rules for the functions an oligosaccharide may have on a glycoprotein, since these depend on the type of glycoprotein, the structure of the oligosaccharide, and the biological context.

9.4.1 Role in Protein Folding, Conformational Stability, and Solubility

As has been shown for numerous soluble and membrane-bound glycoproteins, glycosylation contributes to the folding of newly synthesized proteins in the endoplasmic reticulum, helps to stabilize the conformation of glycoproteins once the proteins have folded properly, and are essential to maintain solubility once soluble glycoproteins have been secreted. Moreover, there is ample evidence that glycans may protect glycoproteins from proteolytic cleavage by either masking proteolytic cleavage sites or conferring intramolecular stability to glycoproteins. Although there are many examples of proteins where folding, conformation, and solubility do not depend on glycosylation, these stabilizing and protective roles of glycoprotein glycans seem to be of general importance. As discovered recently, the $Glc_{1-3}Man_9GlcNAc_2$ glycans on newly synthesized glycoproteins play a key role in protein folding in that they interact with two molecular chaperones, calnexin and calreticulin, in the endoplasmic reticulum, providing a mechanism for promoting folding, oligomeric assembly, and quality control for newly synthesized glycoproteins (Helenius et al. 1997). The importance of this folding and quality control mechanism is highlighted by the observation that inhibition of deglucosylation causes a blockade in the secretion of many secretory proteins (Gross et al. 1983). It is tempting to speculate that the fact that except for albumin almost all plasma proteins are *N*-glycosylated is related to this mechanism.

9.4.2 Barrier and Structural Functions

A second type of function is dependent on biophysical properties generated by the oligosaccharides in conjunction with their polypeptide backbones.This function is pivotal in gel-forming mucins that protect epithelial surfaces in the gastrointestinal, respiratory, and reproductive tracts, in mucin regions of cell surface receptors, and in the three-dimensional structure of the extracellular matrix. Mucins are very large glycoproteins, with a molecular mass of up to several million containing 50%–80% of complex carbohydrate per molecular weight. The glycans are *O*-glycosidically attached to the polypeptide, are tightly packed, and protrude from the molecule, resulting in a "bottle brush"-like overall shape. Due to their negative charge and their tight attachment to the protein backbone, the sialylated and sulfated glycans confer a rigid, rod-like structure upon the glycoprotein monomers. The monomers are linked to each other both covalently by disulfide bonds and noncovalently by glycan–glycan interactions to form a highly viscous gel that provides both a protective barrier and a sliding surface on the epithelia (Allan and Garner 1980). Mucin regions are also present on cell surface receptors such as the LDL receptor where they result in the formation of rodlike molecular domains assumed to extend the distance of the ligand binding site from the cell surface (Jentoft 1990).

9.4.3 "On–Off" and "Tuning" Functions

A third type of oligosaccharide function has been termed "on–off" or "tuning," describing that oligosaccharides may modulate the biological activity of various glycoproteins (Varki 1993). Examples of "on–off" functions include the glycosylation of glycoprotein hormones such as human chorionic gonadotropin (HCG) which is required for stimulation of adenylate cyclase, but not for binding of the hormone to its receptor (Sairam 1989), modulation of the binding activity of the neural cell adhesion molecule N-CAM by polysialylation (Roth et al. 1992), and

Table 3. Biological functions of complex carbohydrates: oligosaccharide-receptor interactions

Type of Binding	Ligand	Receptor	Function
Individual glycoproteins to			
Intracellular receptors	Phosphorylated oligomannosidic *N*-glycans of lysosomal hydrolases	Mannose 6-phosphate receptors	Routing of newly synthesized lysosomal enzymes to lysosomes (Kornfeld 1986; von Figura and Hasilik 1986)
Cell surface receptors	Plasma glycoproteins with terminal β Gal	Asialoglycoprotein receptor of hepatocytes	Clearance of plasma glycoproteins (Ashwell and Harford 1982)
Cells and particles to			
Phagocytes	Cells (erythrocytes, lymphocytes, tumor cells) with exposed terminal β Gal/β GalNAc residues	Macrophage Gal/GalN receptor	Clearance of senescent erythrocytes (Lee et al. 1990; Roos et al. 1983)
	Cells (protozoa, yeast) with exposed terminal D-mannose residues	Macrophage Man/Fuc receptor	Phagocytosis of pathogens (Ezekowitz et al. 1991; Stahl 1992)
	Particles with exposed β Gal residues	Surface immobilized C-reactive protein	Opsonization, phagocytosis (Kempka et al. 1990; Köttgen et al. 1992)
Pathogens (viruses, bacteria, protozoa, fungi)	Terminal sialic acids on host cell gangliosides and glycoproteins	Viral hemagglutinins (e.g., influenca virus, rota virus, enterovirus)	Binding of viruses to the host cell: infection (Markwell and Paulson 1980; Rogers et al. 1986)
	Oligosaccharides of host cell gangliosides and glycoproteins	Bacterial adhesion (e.g., *Escherichia coli*, *Neisseria gonorrhoeae*), fungal receptors (e.g., *Cryptococcus neoformans*, *Candida albicans*), receptors of protozoa (e.g., *Plasmodium falciparum*)	Infection of host cells (Ofek and Sharon 1990; Jimenez-Lucho et al. 1990; Orlandi et al. 1992)

the modulation of receptor-binding affinity and biological activity of hematopoietic growth factors (Cebon et al. 1990), among others.

9.4.4 Oligosaccharide–Receptor Interactions

By virtue of their structural diversity, the glycans of both glycoproteins and glycolipids are extremely well suited to serve as recognition structures that interact specifically with different types of receptors that have binding sites for distinct structural oligosaccharide epitopes. Oligosaccharide–receptor interactions have been shown to be pivotal for numerous biological processes (Table 3). These include (a) the sorting of proteins during intracellular trafficking such as transport of newly synthesized lysosomal hydrolases from the Golgi complex to lysosomes (Kornfeld 1986; von Figura and Hasilik 1986) and that of secretory glycoproteins to the apical surface (Scheiffele et al. 1995) and (b) selective binding of oligosaccharides to sugar-specific receptors termed mammalian lectins, which has been shown to mediate the clearance of plasma glycoproteins and senescent erythrocytes from the circulation by binding to two types of galactose receptors exposed on hepatocytes and Kupffer cells, respectively (Ashwell and Harford 1982; Roos et al. 1983). Other receptors, including those with binding specificity for mannose and fucose expressed on macrophages (Stahl 1992; Ezekowitz et al. 1991), a circulating core-specific lectin (Colley et al. 1988), as well as C-reactive protein immobilized onto cell surfaces (Kempka et al. 1990; Köttgen et al. 1992), are assumed to be involved in the clearance of cell debris and pathogens. In contrast, viruses, bacteria, fungi, and protozoa utilize complex carbohydrates exposed on the surface of host cells for binding, i.e., for the first step in infection (for examples see Table 3).

9.5 Carbohydrate-Mediated Cell Adhesion

Carbohydrate–lectin interactions mediated by selectins and their ligands are pivotal for the adhesion of leukocytes and platelets to endothelial cells under shear stress that initiates the recruitment of leukocytes to sites of acute and chronic inflammation (for selected reviews, see Lasky 1992; Bevilacqua and Nelson 1993; Kansas 1996).

9.5.1 Selectin–Carbohydrate Interactions

The carbohydrate-binding selectin family of adhesion molecules comprises the three related type I transmembrane cell surface proteins L-, E- and P-selectin, all of which are encoded by genes which are in close proximity to each other on human (and mouse) chromosome 1. L-selectin is expressed by leukocytes, while E-selectin is expressed by chronically activated endothelial cells and P-selectin by acutely activated endothelium and platelets. All three selectins have an N-terminal lectin domain which is capable of carbohydrate binding in a Ca^{2+}-dependent fashion, followed by an EGF homology domain and a variable number of tandem short consensus repeat domains homologous to domains found in various complement-regulating proteins. The carbohydrate–selectin interaction tethers flowing leukocytes to stationary cells of the vessel wall, which is comprised mainly of activated or specialized endothelium but also arrested leukocytes and platelets (Ley et al. 1991, 1993). This results in a rapid loss of relative velocity, which is a prerequisite for an ensuing definitive leukocyte arrest mediated by protein–protein interactions between members of the immunoglobulin and integrin superfamilies (von Andrian et al. 1992). In addition, the continuous recirculation of lymphocytes through the blood into peripheral lymph nodes is completely dependent on L-selectin (L-selectin has therefore been referred to as "peripheral lymph node homing receptor"). Lymphocyte homing to mucosa-associated lymphatic tissue involves L-selectin and α4-integrins in an overlapping rather than sequential fashion.

9.5.2 Carbohydrate Selectin Ligands

All selectin-binding molecules contain carbohydrates bearing a strong negative charge due to modification by neuraminic acid, sulfate, or phosphate residues. In vitro, the three selectins bind two related fucose-containing acidic oligosaccharides, sialyl Lewis X (NeuAc α2–3Galβ1-4(Fucα1–3)GlcNAc; CD15s] (Fig. 2) and sialyl Lewis A [NeuAc α2–3Galβ1-3(Fucα1–4)GlcNAc] (Foxall et al. 1992). Clustering of the carbohydrate moieties appears to be important for high-avidity binding to selectins: Tetravalent sialyl Lewis X is much more efficient in blocking lymphocyte adherence to activated endothelium than di- or trivalent sialyl Lewis X (Turunen et al. 1995). Murine lymph node endothelium L-selectin ligands contain sulfated sialyl Lewis X derivatives (Hemmerich et al. 1994) but the relative importance of 6-sulfo sialyl Lewis X (sulfate→6GlcNAc) and 6'sulfo sialyl Lewis X (sulfate→^{6}Gal) remains controversial.

9.5.3 Glycoprotein Selectin Ligands

Selectin-binding complex carbohydrates are presented by various protein scaffolds. With the notable exception of E-selectin ligand 1 (ESL-1), all selectin-binding glycoproteins with known amino acid sequence of the protein core (GlyCAM-1, CD34, MAdCAM-1 und PSGL-1) belong to the sialomucin family, characterized by copious *O*-linked carbohydrate side chains (Shimizu and Shaw 1993). GlyCAM-1 is a secreted L-selectin-binding sialomucin isolated from murine lymph node endothelial cells which bears 6- and 6'-sulfated sialyl Lewis X and which may act as a soluble L-selectin antagonist (Hemmerich et al. 1995). Lymph node endothelial cells contain another L-selectin binding sialomucin, CD34, which has a protein core identical to that found in hematopoietic stem cells. Due to differential glycosylation, only lymph node endothelial cell-derived CD34 binds to L-selectin. The sialomucin MAdCAM-1 also displays L-selectin binding capacity but only when isolated from mesenterial lymph nodes and not from endothelioma cells. Of the sialomucins mentioned, only CD34 has been shown to be also present in human peripheral endothelium, but it remains unclear whether it has any role in mediating leukocyte extravasation into areas of inflammation. The leukocytes themselves may bind all three selectins: ESL-1 binds E-selectin via *N*-linked carbohydrate moieties (Steegmaier et al. 1995), CD162 binds P- and L-selectin (PSGL-1=P-selectin glycoprotein ligand-1) (Guyer et al. 1996). Leukocyte–leukocyte binding mediated by PSGL-1–L-selectin interaction is pivotal in amplifying leukocyte recruitment in a snowball-like manner after the first leukocytes have adhered to the vessel wall (Walcheck et al. 1996). In addition to glycoprotein ligands, selectins may also bind to distinct glycosaminoglycans. Cultivated endothelial cells contain L-selectin-binding heparansulfate proteoglycans (Norgard-Sumnicht et al. 1993). The relevance of these binding sites may be inferred from the observation that heparinase treatment of activated arterial endothelial cells reduces L-selectin-dependent monocyte binding by 80% (Giuffré et al. 1987). Binding of L-selectin chimeras to sialyl Lewis X can be effectively inhibited by ΔUA2Sα1–4GlcNS6Sα1–4IdoA2Sα1–4GlcNS6S and related heparin tetrasaccharides. Intravenous administration of these heparin tetrasaccharides reduces extravasation of neutrophils into areas of acute inflammation similar to the way anti-L-selectin monoclonal antibodies do. As a corrollary, certain batches of heparin exert anti-inflammatory effects in animal models of transplant rejection, adjuvant arthritis, or experimental autoimmune encephalitis (Lider et al. 1989). The anti-inflammatory effects presumably reside in the ability to interfere with leukocyte extravasation but are unrelated to the anticoagulant properties of the heparin batches used.

9.5.4 Therapeutic Inhibition of Carbohydrate–Selectin Interaction

Aberrant or overshooting leukocyte extravasation may do more harm than good. Therefore, inhibition of leukocyte extravasation has been attempted in animal models of various diseases by means of monoclonal antibodies, soluble selectin chimeras, bacterial or synthetically produced lectin-domain-analogous peptides, sialyl Lewis X glycoconjugates, sulfated polysaccharides such as dextran sulfate or fucoidin (which binds both L- and P-selectin), glycyrrhetin-C-fucoside, and chemically synthesized saccharomimetics such as inositol polyphosphates (for review, see Parekh and Edge 1994). The three disorders studied most extensively are postischemic reperfusion injury, acute respiratory distress syndrome (ARDS), and acute meningoencephalitis.

9.5.4.1 Postischemic Reperfusion Injury

Transient ischemia followed by reperfusion elicits inflammatory changes both locally and in the lung. The associated tissue damage in both areas can be reduced by intravenous administration of selectin-blocking monoclonal antibodies, sialyl Lewis X oligosaccharides or fucoidin (Seekamp et al. 1994). For example, myocardial necrosis in cats subjected to coronary ischemia and reperfusion can be reduced by as much as 50% by sialyl Lewis X oligosaccharides or anti-L-selectin monoclonal antibodies (Buerke et al. 1994).

9.5.4.2 ARDS

Selectin inhibition also yields favorable effects in ARDS-like pulmonary damage induced by experimental sepsis, cobra venom factor, or immune complex formation in that L-selectin antagonists effectively reduce leukocyte infiltration, plasma exudation, and pulmonary hemorrhage (Mulligan et al. 1993).

9.5.4.3 Meningoencephalitis

As shown by experiments with cyclophosphamide-induced neutropenia, neutrophils play no major role in the process of eliminating bacteria from the central nervous system during acute meningoencephalitis. On the contrary, reducing leukocyte extravasation into the brain has a favorable impact on the sequelae of bacterial meningoencephalitis. After intracisternal injection of bacterial polysaccharide toxins, pleocytosis and the accumulation of lactate and serum proteins are ameliorated by intravenous administration of selectin antagonists, such as fucoidin or selectin-binding prokaryotic peptides (Angstwurm et al. 1995).

In conclusion, selectin-mediated leukocyte extravasation is pivotal in acute inflammation, appears to occur in chronic inflammation, and may be important for transplant rejection and autoimmune diseases. However, animal data for these situations are scarce in comparison with those for acute neutrophil-mediated disorders. Fortunately, even prolonged selectin–carbohydrate inhibition appears to have little impact on the ability to fend off viral or bacterial intruders. Since the role of selectin–carbohydrate binding is limited to conditions of shear, the extravasation of leukocytes at pathologically low flow rates and the function of already extravasated leukocytes at work in the surrounding tissue is not altered by therapeutic inhibition of the selectin–carbohydrate interaction. Therefore, this approach holds promise both for acute and chronic leukocyte-mediated diseases and awaits phased clinical studies.

9.6 Changes of Glycosylation in Disease

Owing to the improved techniques and methods for analyzing oligosaccharide structures and functions and for unraveling the biosynthetic pathways of glycoprotein glycans, an increasing number of inherited and acquired defects in glycosylation have been described. Reflecting the diverse biological functions of protein glycans, defects in glycosylation may have highly variable consequences in the diseased organism, ranging from lethal aberrations to almost undetectable morphological and functional disorders. Congenital diseases resulting from aberrant protein glycosylation are very rare in humans. These include the lysosomal storage disorders mucolipidosis II (I-cell disease) and mucolipidosis III (Pseudo-Hurler polydystrophy) (Kornfeld 1986; von Figura and Hasilik 1986), congenital dyserythropoietic anemia type II (Fukuda 1991), paroxysmal nocturnal hemoglobinuria (Rosse 1990), carbohydrate-deficient glycoprotein syndrome (Jaeken et al. 1993), and leukocyte adhesion deficiency II (Etzioni et al. 1992), among others. Children suffering from leukocyte adhesion deficiency II are unable to synthesize the sialyl Lewis X and A ligand of E-selectin, resulting in a severe impairment of neutrophil extravasation and, as a consequence, in recurrent severe infections. As compared to primary congenital disorders acquired secondary aberrations in protein glycosylation are more frequent. Patients with rheumatoid arthritis (Parekh et al. 1985), tuberculosis and Crohn's disease (Parekh et al. 1989) have serum IgG molecules with a higher percentage of oligosaccharides on the Fc part which lack terminal galactose residues, resulting in the exposure of terminal N-acetylglucosamine. Decreased galactosylation of IgG presumedly results from a reduced activity of B-cell βF255-galactosyltransferase (Axford et al. 1987); this has been suggested to be of relevance for the pathogenesis of rheumatoid arthritis (Axford 1991) and may be an aid in differential diagnosis (Young et al. 1991). Chronic alcohol consumption causes elevation of serum transferrin isoforms termed carbohydrate-deficient transferrin (CDT), in which the terminal trisaccharides are deficient (Stibler 1991). Incomplete glycosylation of CDT is presumedly caused by a depression in the activities of glycosyltransferases involved in the biosynthesis of the transferrin oligosaccharides (Xin et al. 1995). CDT was proven to be a sensitive marker of alcohol abuse (Stibler 1991). Chronic liver disease has been shown to be associated with an

Table 4. Targeting of drugs, genes and contrast media to the liver

Drug targeting	Gene targeting	Imaging
Anti-neoplastic drugs	Plasmids (Stankovics et al. 1994)	Scintigraphy (Kudo et al. 1992)
Anti-viral drugs (Biessen et al. 1994)	Antisense oligonucleotides	Magnetic resonance imaging (Reimer et al. 1991a,b)
Lipid-lowering drugs		
Tri-antennary Gal-glycans	Asialoorosomucoid-poly-lysine	^{99m}Tc-DTPA-galactosyl HSA
Lactosylated poly-L-lysine	Asialo-alpha-1-acid glycoprotein	Arabino galactan/ pyrrolidinoxyl radicals
Gal-carboxymethyl-dextran		
Lactosaminated HSA		Asialofetuin iron oxid nanoparticles
(2-Hydroxypropyl)-methacrylamide Copolymers-Galoxide		Arabino galactan iron nanoparticles

DTPA, diethylenetriamine–penta acetic acid; HSA, human serum albumin

elevated sialylation of fibrinogen (Martinez et al. 1978). Furthermore, acute and chronic inflammation are associated with specific alterations in the glycosylation of acute-phase proteins suggested to represent a humoral feedback response of the hepatic acute-phase reaction to dampen the selectin-mediated cellular inflammatory reaction (van Dijk et al. 1994). During the past three decades numerous studies have demonstrated that malignant transformation is associated with structural alterations of the carbohydrate groups of glycoproteins and glycolipids. Structural changes of cell surface carbohydrates are exhibited by essentially all experimental and naturally occurring malignomas regardless of type, cause, or stage of development (for review, see Warren et al. 1978; Hakomori 1989; Reading et al. 1988). Structural changes include an elevation of tri- and tetra-antennary *N*-linked glycans, increased poly-*N*–acetyllactosamines, and "bisecting" *N*-acetyl-D-glucosamine residues, the neo-expression of Lewis antigens, and increased sialylation and fucosylation (selected references: Ogata et al. 1976; Yamashita et al.

1984; Kim et al. 1986; Vischer and Reutter 1978). Tumor-associated carbohydrate structures have been shown to be related to metastasis and to tumor immunology (Takano et al. 1994). For example, the expression of increased amounts of sialyl Lewis X and sialyl Lewis A structures known to be ligands for E-selectin expressed on endothelial cells has been suggested to be involved in the attachment of blooD-borne metastatic tumor cells to endothelial cells and their subsequent extravasation (Sawada et al. 1993). This assumption is supported by the observation that inhibitors of N-glycan processing such as swainsonine and castanospermine inhibit metastasis of experimental tumor cells (Dennis 1986). With respect to immunological mechanisms, distinct carbohydrate antigens have been shown to suppress natural killer cell cytotoxicity for tumor cells, thus contributing to the poor prognosis of tumors expressing these antigenic epitopes (for review, see Muramatsu 1993). Anti-carbohydrate antibodies and lectins have been shown to be diagnostically useful to phenotype tumor cells or to monitor serum levels of tumor-associated antigens (for review, see Reading et al. 1988; Muramatsu 1993).

9.7 Carbohydrates as Ligands for Targeting and Imaging

Distinct types of cells, including macrophages, hepatocytes, endothelial cells, and leukocytes express on their surface nonenzymatic, sugar-binding proteins termed lectins, which mediate endocytosis, phagocytosis, and cell adhesion, among other biological functions (see above). The specific binding of carbohydrate ligands to lectins has been utilized to specifically target various types of cargo to these cells. In particular, two galactose-specific receptors, i.e., the asialoglycoprotein receptor that is expressed on hepatocytes and internalizes glycoproteins and small particles exposing galactose, and the galactose-particle receptor on Kupffer cells with preferential binding affinity for large galactose-exposing particles, have been used for targeting drugs, antisense oligodeoxynucleotides, plasmids, and contrast media to the liver (Table 4). For example, lactosylated poly-L-lysine has been successfully used as a carrier for targeting anti-viral drugs to liver parenchymal cells (Biessen et al. 1994), and asialoglycoprotein/polylysine/DNA complexes were employed for in vivo gene delivery to the liver (Stankovics et al. 1994).

Based on the same strategy, imaging of the asialoglycoprotein receptor with technetium-99m galactosyl human serum albumin (Kudo et al. 1992) or with arabinogalactan-coated ultrasmall superparamagnetic iron oxide (Reimer et al. 1991a) was used to quantitatively assess liver function. Moreover, magnetic resonance asialoglycoprotein receptor imaging using arabinogalactan-coated ultrasmall superparamagnetic iron oxide was shown to be suitable to differentiate between primary liver tumors and functional liver tissue (Reimer et al. 1991b). Targeting to the reticuloendothelial system in liver, spleen, and bone marrow has been attempted using carriers exposing D-mannose and L-fucose residues (Gonsho et al. 1994) or by engineering a mannose glycoform of the proteins to be delivered (Brady et al. 1994).

Acknowledgements. Work in the authors' laboratories has been supported by the Deutsche Forschungsgemeinschaft, the BMBF, and the Sonnenfeld-Stiftung. The authors are indebted to M. May and B. Graef for expert secretarial assistance.

References

Abeijon C , Hirschberg CB (1992) Topography of glycosylation reactions in the endoplasmic reticulum. Trends Biochem Sci 17:32–36

Allan A, Garner A (1980) Mucus and bicarbonate secretion in the stomach and their possible role in mucosal protection Gut 21:249–262

von Andrian UH, Hansell P, Chambers JD, Berger EM, Filho IT, Butcher EC, Arfors KE (1992) L-selectin function is required for β2-integrin-mediated neutrophil adhesion at physiological shear rates in vivo. Am J Physiol 263:H1034-1044

Angstwurm K, Weber JR, Segert A, Bürger W, Weih M, Freyer D, Einhäupl KM, DirnagL-U (1995) Fucoidin, a polysaccharide inhibiting leukocyte rolling, attenuates inflammatory responses in experimental pneumococcal meningitis in rats. Neurosci Lett 191:1–4

Ashwell G, Harford J (1982) Carbohydrate-specific receptors of the liver. Annu Rev Biochem 51:531–554

Axford JS, Mackenzie L, Lydyard PM, Hay FC, Isenberg DA, Roitt IM (1987) Reduced B-cell galactosyltransferase activity in rheumatoid arthritis. Lancet 1987, 2:1486–1488

Axford, JS (1991) Oligosaccharides: an optional extra or of relevance to diseases mechanisms in rheumatology? J Rheumatol 18:1124–1127

Bevilacqua MP, Nelson RM (1993) Selectins. J Clin Invest 91:379–387
Biessen EA, Beuting DM, Vietsch H, Bijsterbosch MK, Van Berkel TJ (1994) Specific targeting of the antiviral drug 5-iodo 2'-deoxyuridine to the parenchymal liver cell using lactosylated poly-L-lysine. J Hepatol 21:806–815
Brady RO, Murray GJ, Barton NW (1994) Modifying exogenous glucocerebrosidase for effective replacement therapy in Gaucher disease. J Inherit Metab Dis 17:510–519
Brockhausen I (1995) Biosynthesis of O-glycans of the N-actylgalactosamine-α-Ser/Thr linkage type. In: Montreuil J, Vliegenthart JFG, Schachter H (eds) Glycoproteins. Elsevier, Amsterdam, pp 201–259
Buerke M, Weyrich AS, Zheng Z, Gaeta FCA, Forrest MJ, Lefer AM (1994) Sialyl LewisX-containing oligosaccharide attenuates myocardial reperfusion injury in cats. J Clin Invest 93:1140–1148
Cebon J, Nicola N, Ward M, Gardner I, Dempsey P, Layton J, Duhrsen U, Burgess AW, Nice E, Morstyn G (1990) Granulocyte-macrophage colony stimulating factor from human lymphocytes. The effect of glycosylation on receptor binding and biological activity. J Biol Chem 265:4483–4491
Colley KJ, Beranek MC, Baenziger JU (1988) Purification and characterization of the core-specific lectin from human serum and liver. Biochem J 256:61–68
Cross GA (1990) Glycolipid anchoring of plasma membrane proteins. Annu Rev Cell Biol 6:1–39
Dennis JW (1986) Effects of swainsonine and polyinosinic: polycytidylic acid on murine tumor cell growth and metastasis. Cancer Res 46:5131–5136
Etzioni A, Frydman M, Pollack S, Avidor I, Phillips ML, Paulson JC, Gershoni-Baruch R (1992) Brief report: recurrent severe infections caused by a novel leukocyte adhesion deficiency. New Engl J Med 327:1789–1792
Ezekowitz RAB, Williams DJ, Koziel H, Armstrong MYK, Warner A, Richards FF, Rose RM (1991) Uptake of Pneumocystis carinii mediated by the macrophage mannose receptor. Nature 351:155–158
Ferguson MAJ, Williams AF (1988) Cell surface anchoring of proteins via glycosylpholphatidylinositol structures. Annu Rev Biochem 57:285–320
Figura K von, Hasilik A (1986) Lysosomal enzymes and their receptors. Annu Rev Biochem 55:167–193
Foxall C, Watson SR, Dowbenko D, Fennie C, Lasky LA, Kiso M, Hasegawa A, Asa D, Brandley BK (1992) The three members of the selectin receptor familiy recognize a common carbohydrate epitope, the sialyl LewisX oligosaccharide. J Cell Biol 117:895–902
Fukuda MN (1991) HEMPAS disease: genetic defect of glycosylation. Glycobiology 1:9–16
Fukuda M, Hindsgaul O (1994) Molecular glycobiology. IRL, Oxford

Furukawa K, Kobata A (1991) IgG galactosylation – its biological significance and pathology. Mol Immunol 28:1333–1340
Gasnier F, Rousson R, Lerme F, Vaganay E, Louisot P, Gateau-Roesch O (1992) Mitochondrial dolichyl-phosphate mannose synthase. Purification and immunogold localization by electron microscopy. Eur J Biochem 206:853–858
Giuffré L, Cordey A-S, Monai N, Tardy Y, Schapira M, Spertini O (1997) Monocyte adhesion to activated aortic endothelium: role of L-selectin and heparan sulfate proteoglycans. J Cell Biol 136:945–956
Glick MC (1995) Genetic regulation of terminal glycosylation. In: Montreuil J, Vliegenthart JFG, Schachter H (eds) Glycoproteins. Elsevier, Amsterdam, pp 261–280
Gonsho A, Irie K, Susaki, H, Iwasawa H, Okumo S, Sugawara T (1994) Tissue targeting ability of saccharide-poly(L-lysine) conjugates. Biol Pharm Bull 17:275–282
Green ED, Boime I, Baenziger JU (1986) Biosynthesis of sulfated asparagine-linked oligosaccharides on bovine lutropin. J Biol Chem 261:16309–16316
Gross V, Andus T, Tran-Thi TA, Schwarz RT, Decker K, Heinrich PC (1983) 1-Deoxynojirimycin impairs oligosaccharide processing of alpha 1-proteinase inhibitor and inhibits its secretion in primary cultures of rat hepatocytes. J Biol Chem 258:12203–12209
Guyer DA, Moore KL, Lynam EB, Schammel CM, Rogelj S, McEver RP, Sklar LA (1996) P-selectin glycoprotein ligand-1 (PSGL-1) is a ligand for L-selectin in neutrophil aggregation. Blood 88:2415–2421
Hakomori S (1989) Aberrant glycosylation in tumors and tumor-associated carbohydrate antigens. Adv Cancer Res 52:257–331
Hakomori S (1990) Bifunctional role of glycosphinogolipids. Modulators for transmembrane signaling and mediators for cellular interactions. J Biol Chem 265:18713–18716
Hannun YA, Bell RM (1989) Function of sphingolipids and sphinglipid breakdown products in cellular regulation. Science 243:500–507
Harris RJ, Spellman MW (1993) O-Linked fucose and other post-translational modifications unique to EGF modules. Glycobiology 3:219–224
Hart GW, Haltiwanger RS, Holt GD, Kelly WG (1989) Glycosylation in the nucleus and cytoplasm. Annu Rev Biochem 58:841–874
Hayes BK, Varki A (1993) The biosynthesis of oligosaccharides in intact Golgi preparations from rat liver. Analysis of N-linked and O-linked glycans labeled by UDP-6–3HN-acetylgalactosamine. J Biol Chem 268:16170–16178
Helenius A, Trombetta ES, Hebert DN, Simons JF (1997) Calnexin, calreticulin and the folding on glycoproteins. Trends Cell Biol 7:193–200
Hemmerich S, Bertozzi CR, Leffler H, Rosen SD (1994) Identification of the sulfated monosaccharides of GlyCAM-1, an endothelial-derived ligand for L-selectin. Biochemistry 33:4820–4829

Hemmerich S, Leffler H, Rosen SD (1995) Structure of the O-glycans in GlyCAM-1, an endothelial-derived ligand for L-selectin. J Biol Chem 270:12035–12047
Hemming FW (1995) The coenzymic role of phosphodolichols. In Montreuil J, Vliegenthart JFG, Schachter H (eds) Glycoproteins. Elsevier, Amsterdam, pp 127–143
Hiscock DR, Yanagishita M, Hascall VC (1994) Nuclear localization of glycosaminoglycans in rat ovarian granulosa cells. J Biol Chem 269:4539–4546
Homans SW (1995) Three dimensional structure of oligosaccharides explored by NMR and computer calculations. In: Montreuil J, Vliegenthart JFG, Schachter H (1995) Glycoproteins. Elsevier, Amsterdam, pp 67–86
Hubbard SC (1988) Regulation of glycosylation. The influence of protein structure on N-linked oligosaccharide processing. J Biol Chem 263:19303–19317
Jaecken J, Carchon H, Stibler H (1993) The carbohydrate-deficient glycoprotein syndromes: pre-Golgi and Golgi disorders. Glycobiology 3:423–428
Jentoft N (1990) Why are proteins O-glycosylated? Trends Biochem Sci 15:291–294
Jimenez-Lucho V, Ginsburg V, Krivan HC (1990) Cryptococcus neoformans, Candida albican, and other fungi bind specifically to the glycosphingolipid lactosylceramide (Galβ1-4Glcβ1-1Cer), a possible adhesion receptor for yeasts. Infect Immun 58:2085–2090
Kansas GS (1996) Selectins and their ligands: current concepts and controversies. Blood 88:3259–3287
Kempka G, Roos PH, Kolb-Bachofen V (1990) A membrane-associated form of C-reactive protein is the galactose-specific particle receptor on rat liver macrophages. J Immunol 144:1004–1009
Kim YS, Yuan M, Itzkowitz SH, Sun Q, Kaizu T, Palekar A, Trump BF, Hakomori S (1986) Expression of Le^y and extended Le^y blood group-related antigens in human malignant, premalignant, and non malignant colonic tissues. Cancer Res 46:5985–5992
Kjellen L, Lindahl U (1991) Proteoglycans: structures and interactions. Annu Rev Biochem 60:443:475
Kobata A (1992) Structures and functions of the sugar chains of glycoproteins. Eur J Biochem 209:483–501
Köttgen E, Hell B, Kage A, Tauber R (1992) Lectin specificity and binding characteristics of human C-reactive protein. J Immunol 149:445–453
Kornfeld R, Kornfeld S (1985) Assembly of asparagine-linked oligosaccharides. Ann Rev Biochem 54:631–664
Kornfeld S (1986) Trafficking of lysosomal enzymes in normal and disease states. J Clin Invest 77:1–6

Kudo M, Todo A, Ikekubo K, Hino M (1992) Receptor index via hepatic asialoglycoprotein receptor imaging: correlation with chronic hepatocellular damag. Am J Gastroenterol 87:865–870

Lasky LA (1992) Selectins: interpreters of cell-specific carbohydrate information during inflammation. Science 258:964–969

Lee H, Kelm S, Michalski JC, Schauer R (1990) Influence of sialic acids on the galactose recognizing receptor of rat peritoneal macrophages. Biol Chem Hoppe Seyler 371:307–316

Ley K, Gaethgens P, Fennie C, Singer MS, Lasky LA, Rosen SD (1991) Lectin-like cell adhesion molecule 1 mediates leukocyte rolling in mesenteric venules in vivo. Blood 77:2553–2555

Ley K, Linnemann G, Meinen M, Stoolman LM, Gaethgens P (1993) Fucoidin, but not yeast polyphosphomannan PPME, inhibits leukocyte rolling in venules of the rat mesentery. Blood 81:177–185

Lider O, Baharav E, Mekori YA, Miller T, Naparstek Y, Vlodavsky I, Cohen IR (1989) Suppression of experimental autoimmune diseases and prolongation of allograft survival by treatment of animals with low doses of heparins. J Clin Invest 83:752–756

Lis H, Sharon N (1993) Protein glycosylation. Structural and functional aspects. Eur J Biochem 218:1–27

Markwell MA, Paulson JC (1980) Sendai virus utilizes specific sialo oligosaccharides as host cell receptor determinants. Proc Natl Acad Sci USA 77:5693–5697

Martinez J, Palascak JE, Kwasniak D (1978) Abnormal sialic acid content of the dysfibrinogenemia associated with liver disease. J Clin Invest 61:535–538

Montreuil J, Vliegenthart JFG, Schachter H (1995) Glycoproteins. Elsevier, Amsterdam

Mulligan MS, Paulson JC, De Frees S, Zheng Z-L, Lowe JB, Ward PA (1993) Protective effects of oligosaccharides in P-selectin-dependent lung injury. Nature 364:149–151

Muramatsu T (1993) Carbohydrate signals in metastasis and prognosis of human carcinomas. Glycobiology 3:294–296

Nishimura H, Kawabata S, Kisiel W, Hase S, Ikenaka T, Takao T, Shimonishi Y, Iwanaga S (1989) Identification of a disaccharide (Xyl-Glc) and a trisaccharide (Xyl2-Glc) O-glycosidically linked to a serine residue in the first epidermal growth factor-like domain of human factors VII and IX and protein Z and bovine protein Z. J Biol Chem 264:20320–20325

Norgard-Sumnicht K, Varki NM, Varki A (1993) Calcium-dependent heparin-like ligands for L-selectin in non-lymphoid endothelial cells. Science 261:480–483

Ofek I, Sharon N (1990) Adhesins as lectins: specificity and role in infection. Microbiol Immunol 151:91–114
Ogata S, Muramatsu T, Kobata A (1976) New structural characteristics of the large glycopeptides from transformed cells. Nature 259:580–582
Orlandi PA, Klotz FW, Haynes JD (1992) A malaria invasion receptor, the 175-kilodalton erythrocyte binding antigen of Plasmodium falciparum recognizes the terminal Neu5Ac(α2–3)Gal-sequences of glycophorin A. J Cell Biol 116:901–909
Parekh RB, Edge CJ (1994) Selectins – glycoprotein targets for therapeutic intervention in inflammation. Trends Biotechnol 12:339–345
Parekh RB, Dwek RA, Sutton BJ et al. (1985) Association of rheumatoid arthritis and primary osteoarthritis with changes in the galactosylation pattern of total serum IgG. Nature 316:452–457
Parekh RB, Isenberg D, Rook G, Roitt IM, Dwek R, Rademacher TW (1989) A comparative analysis of disease – associated changes in the galactosylation of serum IgG. J Autoimmun 2:101–114
Rademacher TW, Parekh RB, Dwek RA (1988) Glycobiology. Annu Rev Biochem 57:785–838
Reading CL, Hakomori SI, Marcus DM (eds) (1988) Altered glycosylation in tumor cells. Liss, New York
Reimer P, Weissleder R, Lee AS, Büttner S, Wittenberg J, Brady TJ (1991a) Asialoglycoprotein receptor function in benign liver disease: evaluation with MR imaging. Radiology 178:769–774
Reimer P, Weissleder R, Brady TJ, Yeager AE, Baldwin BH, Tennant BC, Wittenberg J (1991b) Experimental hepatocellular carcinoma: MR receptor imaging. Radiology 180:641–645
Rogers GN, Herrler G, Paulson JC, Klenk HD (1986) Influenza C virus uses 9-O-acetyl-N-acetyl-neuraminic acid as a high affinity receptor determinant for attachment to cells. J Biol Chem 261:5947–5951
Roos PH, Kolb-Bachofen V, Schlepper-Schäfer J, Monsigny M, Stockert J, Kolb H (1983) Two galactose-specific receptors in the liver with different function. FEBS Lett 157:253–256
Rosse WF (1990) Phosphatidylinositol-linked proteins and paroxysmal nocturnal hemoglobinuria. Blood 75:1595–1601
Roth J (1995) Compartmentation of glycoprotein biosynthesis. In: Montreuil J, Vliegenthart JFG, Schachter H (eds) Glycoproteins. Elsevier, Amsterdam, pp 287–312
Roth J, Rutishauser U, Troy F (1992) Polysialic acids. Birkhauser, Basel
Roux L, Holojda S. Sundblad G, Freeze HH, Varki A (1988) Sulfated N-linked oligosaccharides in mammalian cells. I. Complex-type chains with sialic acids and O-sulfated esters. J Biol Chem 263:8879–8889

Sairam MR (1989) Role of carbohydrates in glycoprotein hormone signal transduction. FASEB J 3:1915–1926

Sawada R, Lowe JB, Fukuda M (1993) E-Selectin-dependent adhesion efficiency of colonic carcinoma cells is increased by genetic manipulation of their cell surface lysosomal membrane glycoprotein-1 expression levels. J Biol Chem 268:12675–12681

Schachter H (1986) Biosynthetic controls that determine the branching and microheterogeneity of protein-bound oligosaccharides. Biochem Cell Biol 64:163–181

Schachter H (1995) Glycosyltransferases involved in the synthesis of N-glycan antennae. In: Montreuil J, Vliegenthart JFG, Schachter H (eds) Glycoproteins. Elsevier, Amsterdam, pp 153–199

Schachter H, Brockhausen I (1992) The biosynthesis of serine(threonini)-N-acetylgalactosamine-linked carbohydrate moieties. In: Allen HJ, Kisailus EC (eds) Glycoconjugates. Composition, structure and function. Dekker, New York, pp 263–332

Schauer R (1991) Biosynthesis and function of N- and O-substituted sialic acids. Glycobiology 1:449–452

Scheiffele P, Peränen J, Simons K (1995) N-Glycans as apical sorting signals in epithelial cells. Nature 378:96–98

Seekamp A, Till GO, Mulligan MS, Paulson JC, Anderson DC, Miyasaka M, Ward PA (1994) Role of selectins in local and remote tissue injury following ischemia and reperfusion. Am J Pathol 144:592–598

Shimizu Y, Shaw S (1993) Mucins in the mainstream. Nature 366:630–631

Stahl PD (1992) The mannose receptor and other macrophage lectins. Curr Opin Immunol 4:49–52

Stankovics J, Crane AM, Andrews E, Wu CH, Wu GY, Ledley FD (1994) Overexpression of human methylmalonyl CoA mutase in mice after in vivo gene transfer with asialoglycoprotein/polylysine DNA complexes. Hum Gene Ther 5:1095–1104

Steegmaier M, Levinovitz A, Isenmann S, Borges E, Lenter M, Kocher HP, Kleuser B, Vestweber D (1995) The E-selectin-ligand ESL-1 is a variant of a receptor for fibroblast growth factor. Nature 373:615–620

Stibler H (1991) Carbohydrate-deficient transferrin in serum: a new marker of potentially harmful alcohol consumption reviewed. Clin Chem 37:2029–2037

Takano R, Muchmore E, Dennis JW (1994) Sialylation and malignant potential in tumour cell glycosylation mutants. Glycobiology 4:665–674

Tauber R, Volz B, Kreisel W, Loch N, Orberger G, Xu H, Nuck R, Reutter W (1994) Reprocessing of membrane glycoproteins. In: Wieland F, Reutter W (eds) Glyco- and cellbiology. Springer, Berlin Heidelberg New York, pp 119–130

Turner GA (1992) N-Glycosylation of serum proteins in disease and its investigation using lectins. Clin Chim Acta 208:149–171

Turunen JP, Majuri M-J, Seppo A, Tiisala S, Paavonen T, Miyasaka M, Lemström K, Penttilä L, Renkonen O, Renkonen R (1995) De novo expression of endothelial sialyl Lewis^{a} and sialyl Lewis^{x} during cardiac transplant rejection: superior capacity of a tetravalent sialyl Lewis^{x} oligosaccharide in inhibiting L-selectin-dependent lymphocyte adhesion. J Exp Med 182:1133–1142

Van Dijk W, Turner GA, Mackiewicz A (1994) Changes in glycosylation of acute-phase proteins in health and disease: occurrence, regulation and function. Glycosylat Dis 1:5–14

Varki A (1993) Biological roles of oligosaccharides: all of the theories are correct. Glycobiology 3:97–130

Verbert A (1995) From $Glc_3Man_9GlcNAc_2$-protein to $Man_5GlcNAc_2$-rpotein: transfer 'en bloc' and processing. In: Montreuil J, Vliegenthart JFG, Schachter H (eds) Glycoproteins. Elsevier, Amsterdam, pp 145–152

Vischer P, Reutter W (1978) Specific alterations of fucoprotein biosynthesis in the plasma membrane of Morris hepatoma 7777. Eur J Biochem 84:363–368

Vliegenthart JFG, Montreuil J (1995) Primary structure of glycoprotein glycans. In: Montreuil J, Vliegenthart JFG, Schachter H (1995). Glycoproteins. Elsevier, Amsterdam, pp 13–28

Walcheck B, Moore KL, McEver RP, Kishimoto TK (1996) Neutrophil-neutrophil interactions under hydrodynamic shear stress involve L-selectin and PSGL-1. A mechanism that amplifies initial leukocyte accumulation of P-selectin in vitro. J Clin Invest 98:1081–1087

Warren L, Buck CA, Tuszynski GP (1978) Glycopeptide changes and malignant transformation. A possible role for carbohydrate n malignant behaviour. Biochim Biophys Acta 516:97–127

Yamashita K, Ohkura T, Tachibana Y, Takasai S, Kobata A (1984) Comparative study of the oligosaccharides released from baby hamster kidney cells and their polyoma transformant by hydrazinolysis. J Biol Chem 259:10834–10840

Xin Y, Lasker JM, Lieber CS (1995) Serum carbohydrate-deficient transferrin: mechanism of increase after chronic alcohol intake. Hepatology 22:1462–1468.

Young A, Sumar N, Bodman KB, Goyal S, Sinclair H, Roitt I, Isenberg D (1991) Agalactosyl IgG: an acid to differential diagnosis in early synovitis. Arthritis Rheum 34:1425–1429

10 Oligonucleotide Aptamers as Specific Targeting Devices in Diagnostics and Therapy

S. Wölfl, M. Kujau, A. Siebert, and M. Wolters

10.1	Introduction	135
10.2	Targeting	136
10.3	Molecular Recognition	137
10.4	Structural Motifs of Nucleic Acid Molecules	140
10.5	The Selection Process: SELEX	142
10.5.1	Selection Protocol	142
10.5.2	Selection Profile	144
10.5.3	Counter Selection	146
10.5.4	Selection Targets	147
10.6	*In vivo* Application of Aptamers	150
10.6.1	Stability of Aptamers	150
10.6.2	Combination with Imaging Tools or Other Compounds	153
10.6.3	Binding as a Therapeutic Modality	154
10.7	Conclusion	154
References		155

10.1 Introduction

Precise targeting of pathogenic tissues is of extraordinary importance in medical diagnosis and therapy. The generation of monoclonal antibodies and the technical development in the use of these specific molecules has created high expectations in these fields. The large size as well as the antigenic potential are major limitations of the antibody technology.

In vitro selection techniques allow the creation of highly structured oligonucleotides which bind a disease-specific target with up to picomolar affinity. The binding affinities of these so called aptamers are often higher than those of antibodies and discrimination between target-molecules of interest can be programmed in the course of the selection experiment.

Chemical modifications which have been developed in the context of antisense RNA research significantly increases their half life *in vivo*, making nucleic acid aptamers available as diagnostic or therapeutic devices.

This review summarizes the methodologies used to create nucleic acid aptamers and explores the possibilities of how they might be developed for clinical use. In particular, the use of aptamers in the diagnosis and therapy of cancer is addressed.

10.2 Targeting

Pathogenic processes in the human body often involve the malfunction of molecular mechanisms and regulation pathways. In some cases, a molecular mechanism that is involved in the pathogenic development of the tumor in one tissue may be essential for the normal function of another, healthy tissue. In the pathogenic event, some molecules may be found at locations or at concentrations that would not be present in the normally functioning organ. During carcinogenesis, specific molecular markers such as embryonic proteins may be present that are less prevalent outside the tumor tissue. These molecular events can be utilized to target the tumor tissue with diagnostic or therapeutic agents. Discrimination between tumor and normal tissue can be based on changes in unique tumor-specific marker molecules (carcinoembryonic antigen, CEA; α-fetoprotein, AFP; prostate-specific antigen, PSA) or on differences in the concentration of molecules that are particularly abundant in the tumor tissue, clearly exceeding the level found in the healthy organ (vascular epithelial growth factor, VEGF; epithelial growth factor, EGF).

For therapeutic applications, a dual mechanism may be employed to ensure specific enrichment at the location of the pathogenic tissue: a targeting moiety that recognizes a specific extracellular marker coupled to a therapeutically active compound that acts with high specificity at another level of the pathogenic process. Thus, to elevate the targeting specificity, two targets that are not usually found in the same tissue can be addressed simultaneously, resulting in highly selective intervention.

10.3 Molecular Recognition

All targeting approaches require the precise recognition of a disease-specific marker at the molecular level. The more selective a ligand interacts with its target in the body, the more potent a therapeutic effect is and the less likely it is that side effects will occur.

The most readily available molecules for specific recognition of such targets are antibodies (Bodey et al. 1996; Verhoeyen et al. 1991). The known target molecule is used for immunization to induce antibody production. After a high antibody titer is reached, the antibody population is screened for specifically and selectively binding antibodies. Antibody molecules that provide high selectivity and specificity are prepared as monoclonal antibodies, well-characterized devices that recognize the initially selected target (Köhler and Milstein 1975). Although target recognition is specific, antibodies have some intrinsic disadvantages for *in vivo* applications. Mouse-derived antibodies often elicit a vigorous immune response (human anti-mouse antibody response, HAMA). The treatment of solid tumors is hampered further by the fact that, because of the size (approximately 150 kDa) and the hydrostatic pressure of the tumor, only a very small fraction of antibodies passes the blood vessel to reach the tumor tissue.

Over the last few years, antibody engineering techniques have provided various improvements. The framework, the constant region, is humanized to reduce the risk of undesired immune responses (Harris 1995; Mountain and Adair 1992; Sandhu 1992). Furthermore, the size can be reduced to a domain required for specific recognition. These Fab fragments lack the constant part of a full-size antibody, but still contain the target-binding domain (Fig. 1). The molecular weight of Fab fragments is only 45 kDa. For diagnostic and therapeutic applications, even smaller molecules such as single-chain fragments (scFv) with less non-specific activity but mostly lowered affinity can be constructed by employing highly sophisticated genetic engineering techniques.

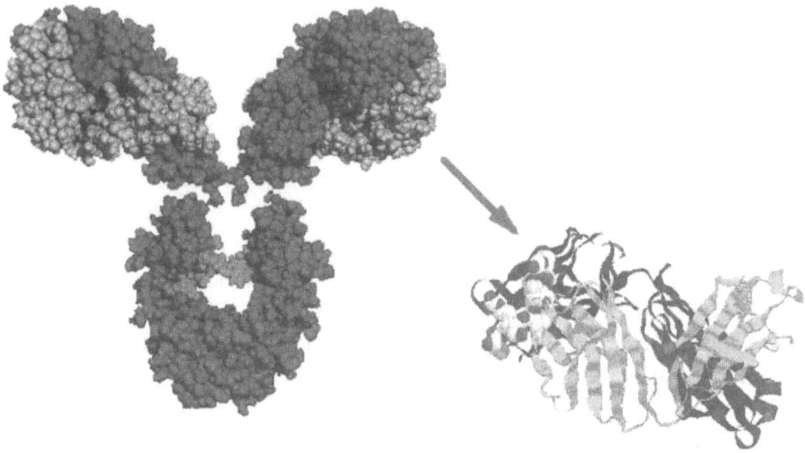

Fig. 1. Monoclonal antibodies are the best-known tools for specific target recognition. *Left*, model of a monoclonal antibody (approximately 150 kDa). Only a small part of the variable domain of the molecule is involved in target recognition. *Right*, the Fab fragment (approximately 45 kDa) still contains the target-specific binding motif (Rini et al. 1993). The figures displayed were obtained from the PDB protein structure database (Abola et al. 1987; Bernstein et al. 1977; Sayle and Milner-White 1995)

In recent years, many methods have been explored that might allow small molecular tools to be developed that will lead to highly specific molecular recognition of a target of interest. In general, these approaches are based on combinatorial chemistry that allows a wide variety of new lead compounds to be synthesized. These strategies include the preparation of various synthetic biological polymers, such as peptides and nucleic acids. Furthermore, low molecular weight compounds can be prepared following the combinatorial approach (Doyle 1995). The system of choice should yield libraries of relatively small molecules that can be prepared so as to differ in structure or chemical properties and allow the simple recovery of specifically and selectively binding molecules. Combinatorial synthesis of small, chemically defined molecules is still very limited with respect to these requirements and is particularly hampered by the problem of identifying a very small number of active molecules.

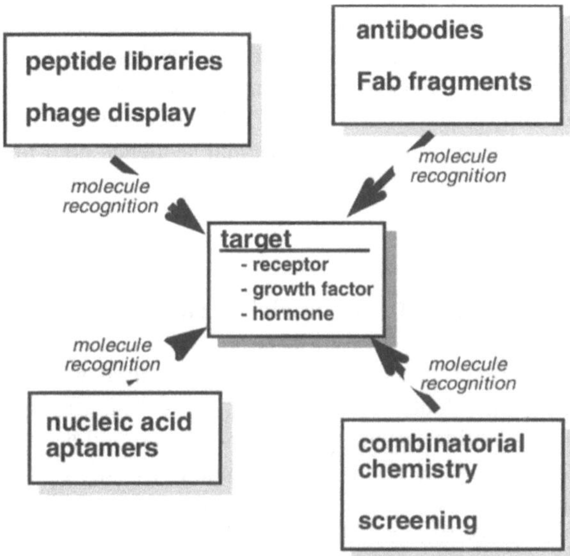

Fig. 2. Various approaches that can be used to generate new molecules for specific target binding. All have in common that from a pool of different structures molecules are selected that specifically recognize the target molecule. Antibody production utilizes natural diversity and selection, whereas the other approaches rely on combinatorial synthesis of a large variety of molecules and subsequent identification *in vitro*

The combination of a simple combinatorial chemical synthesis step with a biological system enables us to fulfill most of the above-mentioned requirements. The various combinatorial approaches for molecular recognition are summarized in Fig. 2. Common to all systems is the preparation of a heterogeneous ensemble of molecules by combinatorial synthesis. In two cases only, molecular biology tools can be used to identify bound molecules after selection: phage/peptide display and *in vitro* selection of nucleic acids. In the following, the combinatorial approach used to create RNA molecules with specific binding affinities (aptamers), their use, and their properties will be described.

By introducing simple combinatorial steps during a single nucleic acid synthesis, a wide variety of molecules with differing sequences are obtained. Molecules of different sequence adopt different structural "phenotypes." As a nucleic acid, its sequence codes for its own replication by appropriate enzymes, providing an individual "genotype" for each molecule present (Gold et al. 1995; Uphoff et al. 1996). Thus, if a binding molecule is present in the original sequence pool obtained by combinatorial synthesis, it can be amplified by using it as a template for its own multiplication.

10.4 Strucutrual Motifs of Nucleic Acid Molecules

The most dominant structure that comes into mind when describing nucleic acids is the double-helix model of DNA. This structure consists of two polynucleotide strands that are wrapped around each other in an antiparallel orientation, are bound horizontally by base pairing and are further stabilized by vertical stacking interactions. The structures formed by single-stranded nucleic acids, however, cannot be reduced to a single structural motif. In fact, many shapes can be adopted. The driving forces are base pairing between two reverse complementary sequence elements and the stacking interaction between adjacent bases. If the proper nucleotide sequence is given, under appropriate conditions only intramolecular structures are formed. The two reverse complementary strands in the double-helical domains are part of the same molecule. The remaining single-stranded regions can adopt a variety of less constrained structures. The double-helical region provides a stabilizing frame, and the scope of recognition motifs is expanded by the variations in the single-stranded domains. Binding can be mediated by hydrogen bonding, stacking of aromatic ring systems, and hydrophobic and electrostatic interactions.

The best-known structural motif of a intramolecular folded nucleic acid is the L-shape of tRNA (Söll and Rajbhandary 1995). Each molecule resembles a good fit for the interacting proteins, such as the appropriate amino acetylases or the tRNA-binding site of the ribosome. Binding to the amino acetylases is highly selective. Only specific tRNAs possessing the correct binding surface are charged with the corresponding amino acid. The fact that a variety of highly specific structures can be formed by single-stranded nucleic acids became even more obvious when enzymatically active RNAs were found. The activity of these

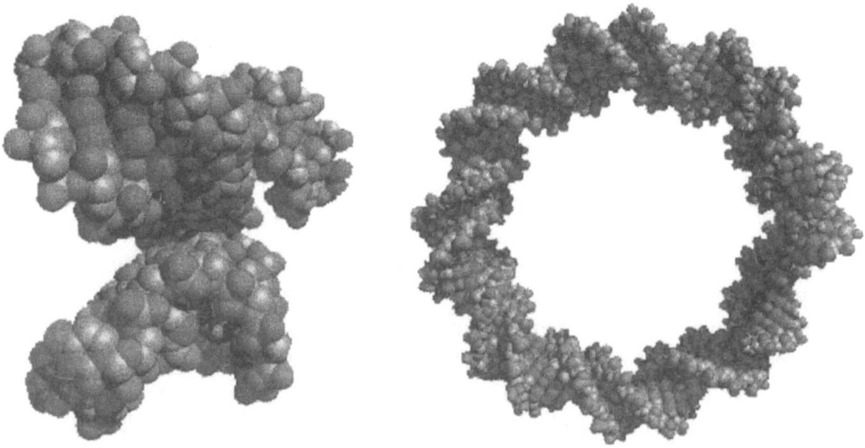

Fig. 3. Nucleic acid molecules can form a variety of structures. Two different models of nucleic acid structures are shown. *Left*, the single-stranded nucleic acid of a ribozyme is folded and forms a stable, defined structure (Tuschl et al. 1994, 1995). *Right*, the most prevalent shape of double-stranded DNA molecules is the well-known regular, repeated shape of the B-DNA double helix (Sussman and Trifonov 1978). The pictures were obtained from the PDB protein structure database (Abola et al. 1987; Bernstein et al. 1977; Sayle and Milner-White 1995)

ribozymes is based on their structural properties (Tuschl et al. 1994, 1995; Fig. 3).

The shape of nucleic acid molecules is inherent to their sequence, although some sequences may be stable in more than one configuration. Alternatively, the same shape with similar binding properties may be formed by unrelated sequences. Nevertheless, a single-stranded nucleic acid molecule will adopt its optimal shape under specific folding conditions. When the sequence of a nucleic acid molecule that provides the desired binding properties is known, it can be prepared in large quantities by chemical synthesis. All molecules of the same sequence will display the same properties under the same conditions.C255

A strategy designed to find molecules that have the perfect shape to bind a chosen target molecule was developed to exploit this property of nucleic acids (Ellington and Szostak 1990; Robertson and Joyce 1990;

Tuerk and Gold 1990). A simple, randomized synthesis approach combined with the intrinsic property of nucleic acids to carry their own "genetic" information led to a simple selection process for binding molecules.

10.5 The Selection Process: SELEX

To perform *in vitro* selection from a large pool of random sequences (systematic evolution of ligands by exponential enrichment, SELEX), some regions of known sequence must be included in the nucleic acid molecules (Fig. 4). The known flanking elements allow the binding molecules to be recovered. The core of random sequence provides the individual element characteristic for each synthetic template. The design and construction of the initial random pool will largely determine which aptamers can be recovered from a selection experiment. The variables that determine the outcome of a selection from the beginning onward include the type of nucleic acid used for screening, the complexity and size of the pool, and the sequences used in the constant flanking regions. In general, a given aptamer sequence is specific to the type of pool it was selected from, so that a given DNA aptamer most probably will not bind its target when the aptamer is formulated as RNA. So far, random sequences between 30 and 200 nucleotides in length were used (Bartel and Szostak 1993). In theory, a given random core of 30 nucleotides may have as many as 10^{18} different sequences, i.e., more variations than molecules present in the initial synthetic pool. A high-quality pool used for selection probably contains between 10^{12} and 10^{15} different sequences. Since many sequences actually adopt related structures and binding motifs, it is very likely that a pool of this size contains most of the potential shape variations.

10.5.1 Selection Protocol

The basic protocol for *in vitro* selection is outlined in Fig. 4. From synthetic single-stranded DNA templates, double-stranded DNA molecules that include a promoter for T7-RNA polymerase are prepared by polymerase chain reaction (PCR). From these double-stranded DNA

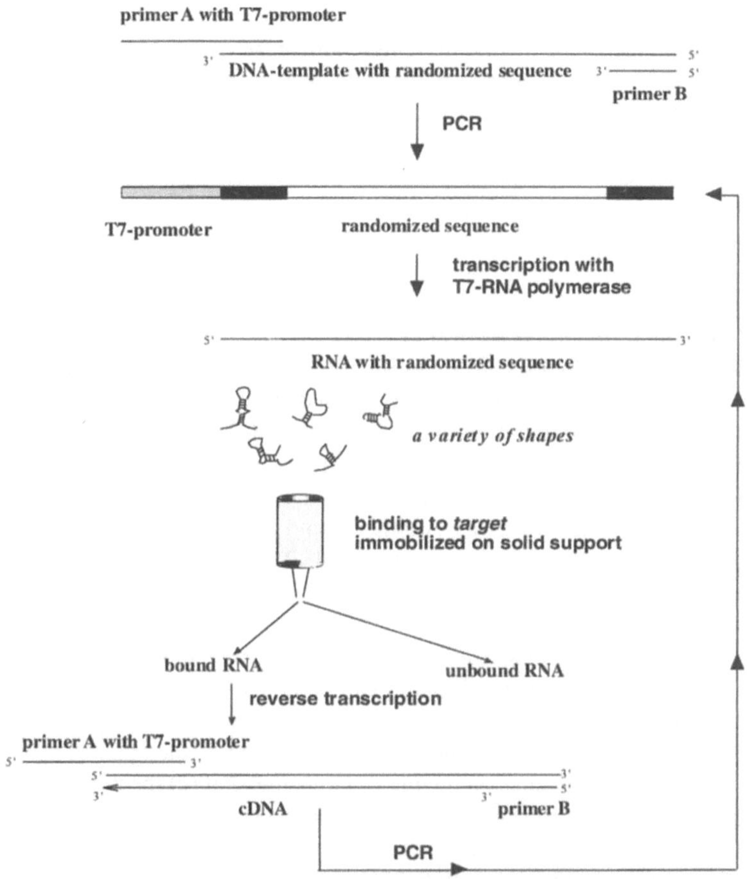

Fig. 4. Selection cycle. The synthetic DNA template contains two features, flanking elements of known sequence and a core element of randomized sequence. Polymerase chain reaction (PCR) and *in vitro* transcription yield a pool of RNA molecules with about 10^{12}–10^{15} different sequences. This pool is incubated with the target molecule. Bound RNA molecules are recovered and copied by reverse transcription. Subsequent PCR and *in vitro* transcription generate a new pool of RNA enriched in binding molecules. Another round of selection adds another degree of binding specificity

templates, a high number of RNA copies are generated by transcription. The resulting RNA pool contains a high number of copies of each of the original 10^{12}– 10^{15} sequence motifs and may include some sequence motifs that will bind the target with various affinities.

For most purposes, the target molecule is immobilized on solid support. Sepharose, magnetic particles, or other activated surfaces are ideally suited to immobilize proteins or other targets. The immobilized target is incubated with the nucleic acid pool (RNA $pool_0$). Alternatively, the target–aptamer complexes can be fractionated by binding to nitrocellulose filters. After binding, unbound molecules are removed and the bound fraction is recovered.

Reverse transcription with a specific primer for the 3' end of the RNA molecule generates cDNA, which is subsequently amplified and converted to double-stranded DNA templates by PCR. This new pool of double-stranded template DNA (dsDNA $pool_{1-n}$) is used to prepare a new pool of RNA (RNA $pool_{1-n}$). This new RNA pool is already enriched in binding molecules. Another round of selection adds another degree of binding specificity to the pool. This selection strategy is repeated for up to 20 cycles until the pool ($pool_n$)contains very tightly binding RNA molecules.

During the course of selection (SELEX), the stringency can be raised in several ways: (a) a natural ligand can be added which competes for target binding; (b) the target concentration can be lowered successively in each round, leading to an increased competition of the aptamers for the available binding sites on the target molecule; and (c) the number of washing steps used to remove unbound molecules can be increased or the buffer composition can be modified in such a way that tighter binding of the aptamer is promoted.

10.5.2 Selection Profile

In the first round, only a small percentage of the molecules of the initial pool will bind to the immobilized target. The initial amount of bound molecules will depend on the selection target, its general affinity for nucleic acids, and the pool used for selection. Figure 5 shows a selection profile observed during several rounds of selection against testosterone immobilized on magnetic particles. The binding of the initial pool was

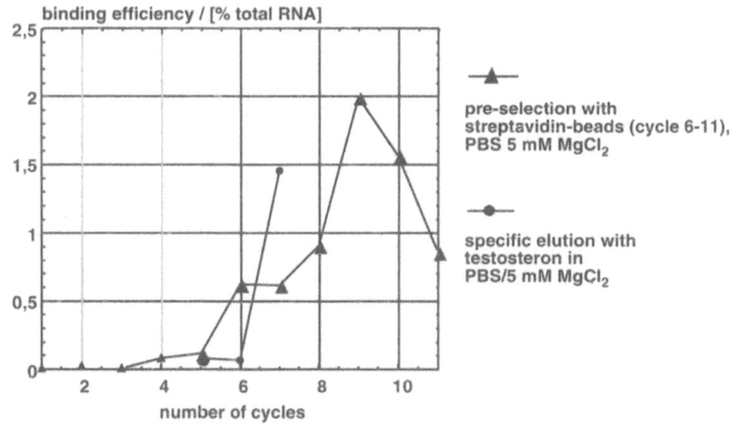

Fig. 5. Example of a selection experiment. An RNA pool containing about 10^{12} different sequence motifs was incubated with testosterone immobilized on magnetic particles. Binding increased significantly in cycles 6–9 (*circles*). After a maximum of binding was reached at cycle 9, binding decreased in subsequent cycles. In parallel, specific elution with free testosterone was carried out in cycles 5–7 (*triangles*)

extremely low. Nevertheless, after five cycles, an increase in binding was observed. The short line from cycle 5 to cycle 8 shows the effect of specific elution with testosterone. Specific elution is one approach that allows molecules that require the matrix for binding to be eliminated. The RNA molecules from the best binding $pool_n$ were converted in the double-stranded DNA form as during the selection cycles. The fragments were cloned and sequenced. Some molecules were selected and analyzed for their binding properties, their binding constants (K_D values), and cross-reactivities. In the presented example, binding constants between 20 and 100 μM were observed.

It is surprising that, beyond a certain amount of cycles, a decrease in binding affinity is observed. This so-called overselection is the result of a different selection pressure. All the enzymatic reactions involved – reverse transcription (RT), PCR, or *in vitro* transcription – have their own sequence preferences. In particular, very stable folding structures tend to interfere with these reactions. Thus, if the binding molecules are not very effective for the required reactions, less structured, weak binding molecules will be enriched during the amplification steps.

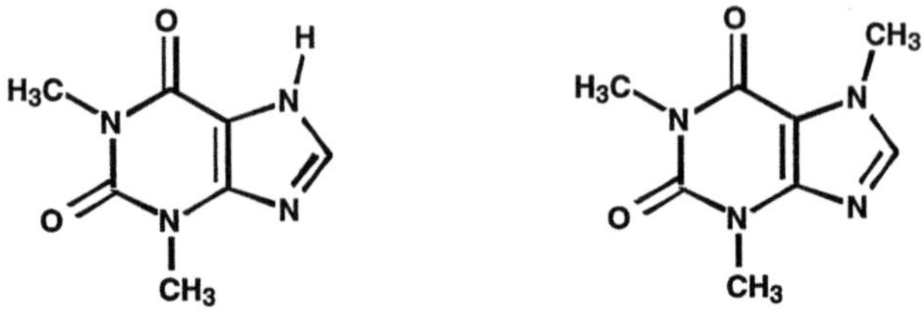

Fig. 6. Theophylline (*left*; K_D, 320 n*M*) and caffeine (*right*; K_D, 3.5 m*M*) are two related molecules that differ in only one methyl group. Jenison et al. (1994) obtained RNA aptamers that specifically recognize theophylline and show almost no affinity for caffeine

10.5.3 Counter Selection

In vitro selection strategies can be thought of as an instrument to investigate molecular targets in terms of the nucleic acid structures they can bind. However, some of these molecules will bind with equal or higher affinity to related molecules or may even bind to the matrix used for immobilization. For example, purine-rich sequences within single-stranded regions tend to bind quite well to cellulose acetate/nitrate filters, which might lead to the amplification of filter-binding parasites. Once such aptamers dominate the oligonucleotide pool, they are extremely difficult to remove from a selection. To ensure the selection of meaningful RNA aptamers, the selection protocol must be intelligently applied so that molecules with undesired affinities are excluded from the binding fraction. This can be achieved by including a negative selection assay. Using several enrichment steps for bound and unbound fractions within one cycle or in separate selection cycles, only molecules binding the expected target will be enriched. Preselection eliminates molecules that bind to the solid phase, while counterselection eliminates molecules that bind to molecules that are similar to the target but contain at least one significant difference. Specificity can be increased additionally by elution of the bound molecules with increased concentrations of free target.

Using the approach of selection and counterselection, Jenison and coworkers selected aptamers that tightly bind to theophylline but do not recognize caffeine (Jenison et al. 1994). The two molecules are bound with a 10000-fold difference in binding affinity. This finding was particularly interesting since theophylline and caffeine differ only in one methyl group (Fig. 6). Because of the small binding surface, the overall affinity of the aptamer is only in the micromolar range. Nevertheless, the oligonucleotide selectively recognizes theophylline even in the presence of large amounts of caffeine.

10.5.4 Selection Targets

One of the major advantages of *in vitro* selection is that there are few limitations concerning the choice of the target. The effectiveness of the selection will depend on the hydrophilic/lipophilic properties of the target. Some molecules may turn out to be almost inaccessible to nucleic acid aptamers. Charged residues on the target molecule are also likely to interfere with nucleic acid binding but, when included in a defined framework, they may also contribute to the specificity of recognition.

Examples of targets used for *in vitro* selection are summarized in Tables 1 and 2. Nucleic acid aptamers that bind to small molecular weight targets are shown in Table 1, while aptamers that recognize larger molecules (proteins or peptides) are listed in Table 2. The latter contains potential targets suitable for diagnostic or therapeutic approaches. Molecules that were selected for their enzymatic activity (ribozymes) are not included. The two tables display a wide variety of molecules. Some of these targets are bound naturally by nucleic acids in the cellular context, and *in vitro* selection is likely to recapitulate or even improve the wild-type binding sequence. Other targets are molecules that are not expected to interact with nucleic acids naturally. However, *in vitro* selection may result in aptamers that will recognize these molecules *in vitro* and *in vivo*. It is important to note that the binding constants indicated in the tables were measured with different techniques and thus are not directly comparable.

Genetically linked selection techniques, such as nucleic acid *in vitro* selection, phage display, or even presentation on bacterial surface proteins, permit many selection strategies to be used. Based on two features – counterselection and reamplification of very small amounts of bound molecules by (RT-)PCR – various selection strategies can be followed that utilize the presence or absence of a distinct feature of a pathogenic

Table 1. Examples of aptamers that recognize low molecular weight ligands

Name of target (ligand)	Type	MW (kDa)	Reference
Cibacron blue 3G-A, reactive blue and other dyes	RNA	< 500 μ*M*	Ellington and Szostak 1990
Cibacron blue 3G-A, reactive blue 4, and other dyes	DNA	ca. 46–33μ*M*	Ellington and Szostak 1992
ATP	RNA	ca. 14 –0.7 μ*M*	Sassanafar and Szostak 1993
Cyanocobalamin	RNA	ca. 10^{-7}	Lorsch and Szostak 1994
Caffeine, theophylline	RNA	ca. 10^{-7}	Jenison et al. 1994
Adenosine/ATP	ssDNA	ca. 10^{-6}	Huizenga and Szostak 1995
Neomycin	RNA	ca. 10^{-7}	Wallis et al. 1995
Aminoglycoside antibiotics	RNA	ca. 10^{-5}	Wang and Rando 1995
l-Arginine,citrulline	RNA	ca. 10^{-5}	Burgstaller et al. 1995
l-Arginine	RNA	ca. 3.3×10^{-7}	Geiger et al. 1996
Porphyrins	DNA		Li et al. 1996
d/l-Adenosine d/l-Arginine	Mirror-image RNA	ca. 1 μ*M*	Klussmann et al. 1996; Nolte et al. 1996

ssDNA, single-stranded DNA.

Table 2. Examples of aptamers that recognize protein and peptide ligands

Name of target (ligand)	Type	MW (kDa)	Reference
T4 DNA polymerase	RNA		Tuerk and Gold 1990
Thrombin	ssDNA	ca. 200–25 n*M*	Bock et al. 1992
HIV Rev protein	RNA		Giver et al. 1993
Human α-thrombin	RNA		Kubik et al. 1994
Protein kinase C	RNA		Conrad et al. 1994
Reverse transcriptase	RNA		Chen and Gold 1994
VEGF	RNA		Jellinek et al. 1994
HNE	Amino-RNA	ca. 30–10 n*M*	Lin et al. 1994
VEGF	Amino-RNA	ca. 2.4 n*M*	Green et al. 1995
Basic FGF	Amino-RNA	ca. 350 n*M*	Jellinek et al. 1995
HIV Rev protein	RNA	ca. 6–0.8 n*M*	Jensen et al. 1995
HIV-1 integrase	RNA	ca. 10^{-6}– 10^{-8}	Allen et al. 1995
L22 ribosomal protein	RNA		Dobbelstein and Schenk 1995
NGF	RNA		Binkley et al. 1995
RT of HIV-1	ssDNA	ca. 10^{-9}	Schneider et al. 1995
Tachykinin substance P	RNA	ca. 10^{-7}	Nieuwlandt et al. 1995
Tax	RNA		Tian et al. 1995
HNE	RNA		Smith et al. 1995
HNE	DNA		Davis et al. 1996
IgE (Fcε receptor I)	Amino-RNA	ca. 30–35 n*M*	Wiegand et al. 1996
IgE (Fcε receptor I)	ssDNA	ca. 10^{-8}	Wiegand et al. 1996
L-Selectin	DNA		O'Connell et al. 1996
RT of FIV	RNA		Chen et al. 1996

HIV, human immunodeficiency virus; VEGF, vascular endothelial growth factor; HNE, human neutrophil elastase; FGF, fibroblast growth factor; NGF, nerve growth factor; RT, reverse transcriptase; FIV, feline immunodeficiency virus; ssDNA, single-stranded DNA.

process. This can be an as yet unknown molecule present in the disease tissue or merely a significant structural difference due to a specific combination of abundant proteins in the disease tissue. In the latter approach, molecules (aptamers) that become enriched in the target tissue are recovered directly from tissue samples. An approach of this type has already been successful with phage epitopes (Pasqualini and Ruoslahti 1996). Instead of presenting a well-defined target *in vitro*, the pool is incubated in the living organism and the molecules that reach the target tissue are amplified and used for another round of selection. Alternatively, patient serum that may contain the elusive marker may be as succesful as direct *in vivo* selection (Tsai and Keene 1993).

A successfully discriminating molecule found *in vivo* may even be used to identify the biologically relevant molecule(s) or mechanism found in the target tissue. In addition, using an *in vivo* targeting approach during selection, a variety of counterselection principles are directly included. Nonspecific binding to other molecules is limited, as *in vivo* stability must be sufficient and clearance slow enough to permit the targeting molecule to reach its target.

Another important feature of aptamers is that their specificity can be changed. A partially randomized pool is prepared from a known aptamer and used for selection against a related molecule. Using this approach, the binding specificity of an l-citrulline-specific aptamer was changed to l-arginine-specific binding (Famulok 1994).

10.6 *In vivo* Application of Aptamers

10.6.1 Stability of Aptamers

Applications in diagnosis or therapy require molecules that have a fairly long half-life in the human body and high bioavailability. The plasma level of the molecule in the human body must be high enough over a wide time range. Within this range, the aptamer must reach the target and exert its action before it is degraded or eliminated from the body. Most natural nucleic acids do not fulfill these criteria. Cleaved by endogenous nucleases, they are readily inactivated and cleared from the human body. For example, the half-life of a short RNA molecule in rat blood is only a few minutes. This is mainly due to the ubiquitous

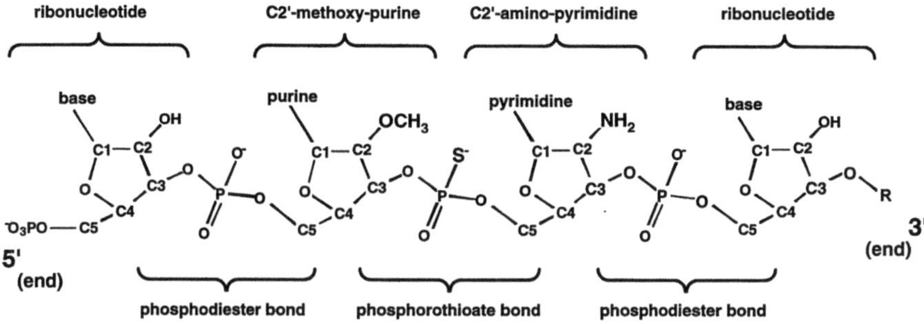

Fig. 7. The sugar phosphate backbone of RNA. In this example, natural ribonucleotides are shown together with modified nucleotides. Modification at the C2' position of the ribose and modification of the phosphate group prevent the cleavage of the RNA molecule with nucleases at these positions. C2'-methoxy-purine, C2'-amino-pyrimidine, and phosphorothioate modifications are shown

presence of nucleases. In recent years, nucleic acids have been chemically modified to alter the structure of an oligonucleotide in such a way that it is no longer the substrate of nucleases. However, these modified nucleic acids must retain the basic properties of nucleic acids. They must be compatible with the enzymatic reactions of the SELEX process, and they must still be capable of folding into sequence-specific structures. Several modifications developed to stabilize oligonucleotides for antisense applications *in vivo* retain most of these nucleic acid properties, in particular specific base pairing, a feature also essential for the shape of aptamere molecules.

The basic approaches used include modification at the phosphodiester bond and substitution of the 2'-hydroxyl group of the ribose (Fig. 7). Unfortunately, most of these modifications have a significant influence on the overall structure of the molecule. Introduced after selection, these modifications are likely to significantly change the binding specificity of the selected aptamer. Therefore, modifications must be used that can be present throughout the selection process, providing all the properties that influence the specific binding of the target. Thus modifications must be compatible with all enzymatic reactions required in the selection protocol.

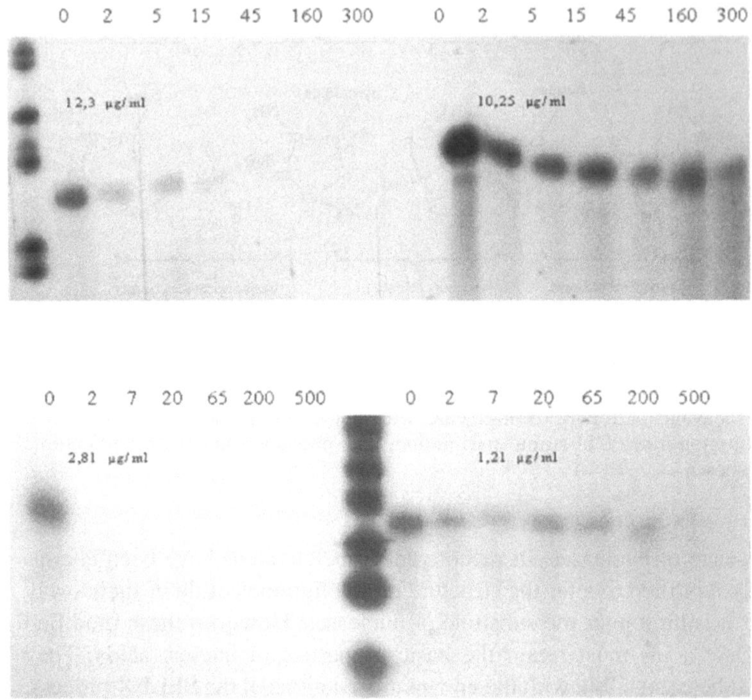

Fig. 8. The stability of nucleic acids in biological fluids was tested in rat blood as a measure for *in vivo* stability. Unmodified RNA (*left*) and 2'-amino-pyrimidine RNA (*right*) were incubated in rat blood for the times indicated. The concentrations per milliliter of rat blood are indicated. Densitometric analysis gave half-lives of less than 5 min for unmodified RNA and of about 8 h for the 2'-amino-pyrimidine-modified molecules

A very successful approach is the substitution of the 2'-hydroxyl groups of the ribose in RNA molecules with amino, fluoro, or *O*-methyl groups (Fig. 7; Heidenreich et al. 1993). It was shown that triphosphates containing these modifications are still substrates of RNA polymerase, but incorporation in the nascent RNA is much slower (Aurup et al. 1994; Lin et al. 1994). Therefore, substitution at all 2'-hydroxyl groups is not feasible. Pyrimidine residues are more sensitive to degradation due to a higher activity of pyrimidine-specific nucleases. Even by using only modified pyrimidine residues, a significant increase in half-life is obtained. Although efficiencies of enzymatic reactions are lower, 2'-amino-pyrimidine RNA can be used throughout the selection process (Lin et al. 1994). The resulting amino-modified RNAs are fairly stable in biologically fluids, making these molecules useful devices for *in vitro* application.

In Fig. 8, a comparison of the stability of 2'-amino-pyrimidine RNA and unmodified RNA in rat blood is shown. The unmodified RNA was immediately degraded and could no longer be detected after 10 min; 2'-amino-pyrimidine-modified molecules were still detectable even after 8 h of incubation.

Stability can be additionally improved by terminal modifications. Modifying the ends of the oligonucleotide aptamer, the cleavage mediated by exonucleases can be blocked. These terminal modifications are also of great interest to provide reactive groups that allow the aptamer molecules to be linked with other moieties. In general, terminal modifications can be introduced after the selection. In most cases, they should not change the structure of the aptamer and will therefore not change its affinity for the selected target.

A new approach was recently demonstrated that yields nucleic acids that are not degraded by ribonucleases (Klussmann et al. 1996; Nolte et al. 1996). First, selection is carried out with unmodified natural nucleic acids against the mirror image of the target. This yields aptamers (d-RNA-aptamer) that bind with high affinity to this mirror image. Based on the sequence of these oligonucleotides, the l-RNA aptamer can be synthesized, yielding a mirror image of the selected molecule. This l-RNA-aptamer ("*spiegelmer*") specifically recognizes the original target. As a stereoisomer of natural d-RNA, it is no longer a substrate for enantioselective nucleases and is therefore not degraded. In human serum, *spiegelmers* could still be detected after 60 h of incubation (Klussmann et al. 1996).

10.6.2 Combination with Imaging Tools or Other Compounds

As long as there is no interference with the folding process, nucleic acid aptamers may be readily coupled with a variety of non-nucleic-acid molecules and still retain their function. Thus signal molecules for imaging or therapeutic effectors are easily combined with the aptamer. During chemical synthesis, a specific substituent at either end of the oligonucleotide can be introduced and used for coupling of the active partner molecule, e.g., a chelator unit that binds ^{99m}Tc for *in vivo* diagnostic imaging. This domain should not interfere with binding to the target. Optimal distances between the aptamer and the active unit must be adjusted using appropriate linker molecules.

These conjugates must be tested to determine whether the activity has been retained. It also should be expected that not all moieties can be combined with equal efficiency. Additional moieties may be coupled to optimize the overall *in vivo* properties of the aptamer. This may include compounds that influence bioavailability or the clearance pathways.

10.6.3 Binding as a Therapeutic Modality

In some cases, aptamers may even be effective without further modification. The aptamer may directly interfere with the binding of the physiological ligand, e.g., inactivating a cell surface receptor. Another aptamer may act as antagonist. If selection was successful and the structure of the binding motif could be analyzed by high-resolution techniques such as X-ray crystallography or nuclear magnetic resonance spectroscopy (NMR), it could be used as a model to rationally design synthetic analogue molecules (mimetics). In this way, new small, active synthetic compounds may become available which most certainly are more cost-effective than nucleic acid oligomers. In another strategy for *intracellular* application, the aptamer sequence could be part of an *in vivo* transcription unit transcribed in the target tissue, thereby producing an aptamer oligonucleotide which is directly released into the cytoplasm.

10.7 Conclusion

Nucleic acid aptamers provide a novel device that promises highly specific and selective molecular recognition. The range of ligands recognized specifically by nucleic acid aptamers includes proteins, such as VEGF, basic fibroblast growth factor (bFGF), RT, and thrombin, and small molecules, such as amino acids, dyes, antibiotics, and even ATP (see Tables 1, 2). By applying counterselection, it is possible to obtain aptamers that clearly discriminate between two related molecules, as shown for theophylline and caffeine (Jenison et al. 1994). Oligonucleotide aptamers for more than 100 protein targets have been developed so far, generally with affinities in the nanomolar range or even better.

,0,0By binding to cell surface proteins, they can target specific cell types and tissues. By binding to diffusible messenger molecules, hormones, growth factors, or signaling molecules, they provide a modulator of signal activity. Aptamers provide model structures that recognize given targets. These lead structures can potentially be used to synthesize even smaller molecules with analogue binding moieties limited to the recognition site.

Chemical synthesis of nucleic acids provides the basis for coupling any aptamer with additional effector moieties that allow imaging or provide additional pharmacological activity, e.g., cell toxicity. Chemical synthesis also guarantees that each aptamer will be prepared with constant highly quality following good manufacturing practice (GMP) guidelines.

The wide range of affinities for molecular targets, the availability of stabilizing modifications, their relatively small size, and the ease of their synthesis make aptamers a device with promising potential for applications in disease management, including diagnostics and therapeutics.

Acknowledgements. We would like to thank Raffaela Santoro for helpful discussion of the manuscript. Anke Siebert is a fellow of the graduate student program of the Friedrich Schiller University in Jena.

References

Abola EE, Bernstein FC, Bryant SH, Koetzle TF, Weng J (1987) Protein data bank. In: Allen FH, Bergerhoff G, Sievers R (eds) Crystallographic databases – information content, software systems, scientific applications. Data Commission of the International Union of Crystallography, Bonn/Cambridge/Chester, pp 107–132

Allen P, Worland S, Gold L (1995) Isolation of high-affinity RNA ligands to HIV-1 integrase from a random pool. Virology 209:327–336

Aurup H, Tuschl T, Benseler F, Ludwig J, Eckstein F (1994) Oligonucleotide duplexes containing 2'-amino-2'-deoxycytidines: thermal stability and chemical reactivity. Nucleic Acids Res 22:20–24

Bartel DP, Szostak JW (1993) Isolation of new ribozymes from a large pool of random sequences. Science 261:1411–1418

Bernstein FC, Koetzle TF, Williams GJB, Meyer JEF, Brice MD, Rodgers MD, Kennard O, Shimanouchi T, Tasumi M (1977) The protein data bank: a computer-based archival file for macromolecular structures. J Mol Biol 112:535–542

Binkley J, Allen P, Brown DM, Green L, Tuerk C, Gold L (1995) RNA ligands to human nerve growth factor. Nucleic Acids Res 23:3198–3205

Bock LC, Griffin LC, Latham JA, Vermaas EH, Toole JJ (1992) Selection of single-stranded DNA molecules that bind and inhibit human thrombin. Nature 355:564–566

Bodey B, Siegel SE, Kaiser HE (1996) Human cancer detection and immunotherapy with conjugated and non-conjugated monoclonal antibodies. Anticancer Res 16:661–674

Burgstaller P, Kochoyan M, Famulok M (1995) Structural probing and damage selection of citrulline- and arginine-specific RNA aptamers identify base positions required for binding. Nucleic Acids Res 23:4769–4776

Chen H, Gold L (1994) Selection of high-affinity RNA ligands to reverse transcriptase: inhibition of cDNA synthesis and RNase H activity. Biochemistry 33:8746–8756

Chen H, McBroom DG, Zhu YQ, Gold L, North TW (1996) Inhibitory RNA ligand to reverse transcriptase from feline immunodeficiency virus. Biochemistry 35:6923–6930

Conrad R, Keranen LM, Ellington AD, Newton AC (1994) Isozyme-specific inhibition of protein kinase C by RNA aptamers. J Biol Chem 269:32051–32054

Davis KA, Abrams B, Lin Y, Jayasena SD (1996) Use of a high affinity DNA ligand in flow cytometry. Nucleic Acids Res 24:702–706

Dobbelstein M, Shenk T (1995) *In vitro* selection of RNA ligands for the ribosomal L22 protein associated with Epstein-Barr virus-expressed RNA by using randomized and cDNA-derived RNA libraries. J Virol 69:8027–8034

Doyle PM (1995) Combinatorial chemistry in the discovery and development of drugs. J Chem Technol Biotechnol 64:317–324

Ellington AD, Szostak JW (1990) *In vitro* selection of RNA molecules that bind specific ligands. Nature 346:818–822

Ellington AD, Szostak JW (1992) Selection *in vitro* of single-stranded DNA molecules that fold into specific ligand-binding structures. Nature 355:850–852

Famulok M (1994) Molecular recognition of amino acids by RNA-Aptamers: an L-citrulline binding RNA motif and its evolution into an arginine binder. J Am Chem Soc 116:1698–1706

Geiger A, Burgstaller P, von der Eltz, H, Roeder A, Famulok M (1996) RNA aptamers that bind L-arginine with sub-micromolar dissociation constants and high enantioselectivity. Nucleic Acids Res 24:1029–1036

Giver L, Bartel D, Zapp M, Pawul A, Green M, Ellington AD (1993) Selective optimization of the Rev-binding element of HIV-1. Nucleic Acids Res 21:5509–5516

Gold L, Polisky B, Uhlenbeck O, Yarus M (1995) Diversity of oligonucleotide functions. Annu Rev Biochem 64:763–797

Green LS, Jellinek D, Bell C, Beebe LA, Feistner BD, Gill SC, Jucker FM, Janjic N (1995) Nuclease-resistant nucleic acid ligands to vascular permeability factor/vascular endothelial growth factor. Chem Biol 2:683–695

Harris WJ (1995) Production of humanized monoclonal antibodies for *in vivo* imaging and therapy. Biochem Soc Trans 23:1035–1038

Heidenreich O, Pieken W, Eckstein F (1993) Chemically modified RNA: approaches and applications. FASEB J 7:90–96

Huizenga DE, Szostak JW (1995) A DNA aptamer that binds adenosine and ATP. Biochemistry 34:656–665

Jellinek D, Green LS, Bell C, Janjic N (1994) Inhibition of receptor binding by high-affinity RNA ligands to vascular endothelial growth factor. Biochemistry 33:10450–10456

Jellinek D, Green LS, Bell C, Lynott CK, Gill N, Vargeese C, Kirschenheuter G, McGee DP, Abesinghe P, Pieken WA et al (1995) Potent 2'-amino-2'-deoxypyrimidine RNA inhibitors of basic fibroblast growth factor. Biochemistry 34:11363–11372

Jenison RD, Gill SC, Pardi A, Polisky B (1994) High-resolution molecular discrimination by RNA. Science 263:1425–1429

Jensen KB, Atkinson BL, Willis MC, Koch TH, Gold L (1995) Using *in vitro* selection to direct the covalent attachment of human immunodeficiency virus type 1 Rev protein to high-affinity RNA ligands. Proc Natl Acad Sci USA 92:12220–12224

Klussmann S, Nolte A, Bald R, Erdmann VA, Fürste JP (1996) Mirror-image RNA that binds D-adenosine. Nat Biotechnol 14:1112–1115

Köhler G, Milstein C (1975) Continuous cultures of fused cells secreting antibody of predefined specificity. Nature 256:495–497

Kubik MF, Stephens AW, Schneider D, Marlar RA, Tasset D (1994) High-affinity RNA ligands to human alpha-thrombin. Nucleic Acids Res 22:2619–2626

Li Y, Geyer CR, Sen D (1996) Recognition of anionic porphyrins by DNA aptamers. Biochemistry 35:6911–6922

Lin Y, Qiu Q, Gill SC, Jayasena SD (1994) Modified RNA sequence pools for *in vitro* selection. Nucleic Acids Res 22:5229–5234

Lorsch JR, Szostak JW (1994) *In vitro* selection of RNA aptamers specific for cyanocobalamin. Biochemistry 33:973–982

Mountain A, Adair JR (1992) Engineering antibodies for therapy. Biotechnol Genet Eng Rev 10:1–142

Nieuwlandt D, Wecker M, Gold L (1995) *In vitro* selection of RNA ligands to substance P. Biochemistry 34:5651–5659

Nolte A, Klussmann S, Bald R, Erdmann VA, Fürste JP (1996) Mirror-design of L-oligonucleotide ligands binding to L-arginine. Nature Biotechnology 14:1116–1119

O'Connell D, Koenig A, Jennings S, Hicke B, Han HL, Fitzwater T, Chang YF, Varki N, Parma D, Varki A (1996) Calcium-dependent oligonucleotide antagonists specific for L-selectin. Proc Natl Acad Sci USA 93:5883–5887

Pasqualini R, Ruoslahti E (1996) Organ targeting *in vivo* using phage display peptide libraries. Nature 380:364–366

Rini JM, Stanfield RL, Stura EA, Salinas PA, Profy AT, Wilson IA (1993) Crystal structure of a human immunodeficiency virus type 1 neutralizing antibody, 50.1, in complex with its V3 loop peptide antigen. Proc Natl Acad Sci USA 90:6325–6329

Robertson DL, Joyce GF (1990) Selection *in vitro* of an RNA enzyme that specifically cleaves single-stranded DNA. Nature 344:467–468

Sandhu JS (1992) Protein engineering of antibodies. Crit Rev Biotechnol 12:437–462

Sassanfar M, Szostak JW (1993) An RNA motif that binds ATP. Nature 364:550–553

Sayle R, Milner-White EJ (1995) RasMol: biomolecular graphics for all. TIBS 20 (9):333–379

Schneider DJ, Feigon J, Hostomsky Z, Gold L (1995) High-affinity ssDNA inhibitors of the reverse transcriptase of type 1 human immunodeficiency virus. Biochemistry 34:9599–9610

Smith D, Kirschenheuter GP, Charlton J, Guidot DM, Repine JE (1995) *In vitro* selection of RNA-based irreversible inhibitors of human neutrophil elastase. Chem Biol 2:741–750

Söll D, Rajbhandary UL (eds) (1995) tRNA: structure, biosynthesis and function. American Society for Microbiology, Washington DC

Sussman JL, Trifonov EN (1978) Possibility of nonkinked packing of DNA in chromatin. Proc Natl Acad Sci USA 75:103–107

Tian Y, Adya N, Wagner S, Giam CZ, Green MR, Ellington AD (1995) Dissecting protein:protein interactions between transcription factors with an RNA aptamer. RNA 1:317–326

Tsai DE, Keene JD (1993) *In vitro* selection of RNA epitopes using autoimmune patient serum. J Immunol 150:1137–1145
Tuerk C, Gold L (1990) Systematic evolution of ligands by exponential enrichment: RNA ligands to bacteriophage T4 DNA polymerase. Science 249:505–510
Tuschl T, Gohlke C, Jovin TM, Westhof E, Eckstein F (1994) A three-dimensional model for the hammerhead ribozyme based on fluorescence measurements. Science 266:785–789
Tuschl T, Thomson JB, Eckstein F (1995) RNA cleavage by small catalytic RNAs. Curr Opin Struct Biol 5:296–302
Uphoff KW, Bell SD, Ellington AD (1996) *In vitro* selection of aptamers: the dearth of pure reason. Curr Opin Struct Biol 6:281–288
Verhoeyen ME, Saunders JA, Broderick EL, Eida SJ, Badley RA (1991) Reshaping human monoclonal antibodies for imaging and therapy. Dis Markers 9:197–203
Wallis MG, Ahsen U, Schroeder R, Famulok M (1995) A novel RNA motif for neomycin recognition. Chem Biol 2:543–552
Wang Y, Rando RR (1995) Specific binding of aminoglycoside antibiotics to RNA. Chem Biol 2:281–290
Wiegand TW, Williams PB, Dreskin SC, Jouvin MH, Kinet JP, Tasset D (1996) High-affinity oligonucleotide ligands to human IgE inhibit binding to Fc epsilon receptor I. J Immunol 157:221–230

11 In Vivo Monitoring of Gene Therapy by Radiotracer Imaging

R.G. Blasberg and J.G. Tjuvajev

11.1	Introduction	162
11.2	Imaging the Expression of Transduced Genes	163
11.3	Potential Clinical Impact	164
11.4	Characteristics of a Good "Marker Gene" and "Marker Substrate"	165
11.5	Selection of HSV1-tk as the Marker Gene	166
11.6	Choice of Marker Substrates for Imaging HSV-TK Expression	167
11.7	Other Marker Substrates for Imaging HSV-TK Expression	179
11.8	In Vitro Experimental Studies	170
11.8.1	Reagents	170
11.8.2	Independent Assays of HSV1-tk Gene Expression in Different RG2TK Cell Lines	171
11.8.3	Uptake of Marker Substrates for Identifying HSV1-tk Gene Expression in Tissue Culture	171
11.8.4	Data Analysis	176
11.8.5	Phosphorylated Metabolites and Incorporation of Radioactivity into DNA	177
11.8.6	Relationship Between FIAU Uptake and Two Independent Measures of HSV1-tk Expression	178
11.9	In Vivo Experimental Studies	180
11.9.1	Animal Preparation and Tumor Inoculation	180
11.9.2	Selective Imaging of HSV1-TK Expression with FIAU in Experimental Animals	180
11.9.3	Imaging HSV1-tk Expression Following Transduction of Preestablished Wild-Type RG2 Tumors	183
11.10	Feasibility of Clinical Applications Using PET and SPECT Imaging	189

11.10.1 Safety Issues 190
11.10.2 Dosimetry 191
11.11 Conclusions 192
Appendix 193
References 197

11.1 Introduction

The increasing knowledge of the genetics and molecular biology of human disease, particularly in human cancer (Culver and Blaese 1994), has led to the application of new biological approaches in therapy. A limitation of many biological-based therapies has been our inability to achieve controlled and effective delivery of biologically active molecules to tumor cells or their surrounding matrix. This condition is particularly limiting for biologically active compounds that have very short half-lives and exhibit site-specific therapeutic and toxic effects. Gene-based therapy can provide control over the level, timing, and duration of action of these biologically active products by including specific promoter/activator elements in the genetic material transferred, resulting in more effective therapeutic interventions. Methods are actively being developed for controlled gene delivery to various somatic tissues and tumors using novel formulations of DNA and for controlling gene expression using cell-specific, replication-activated,and drug-controlled expression systems (Jolly 1994).

Current gene-based therapies for cancer involve several different biological approaches to treatment, and new approaches continue to be developed. There are over 100 human gene therapy clinical trials currently being investigated in the United States (Clinical Protocol 1995), and additional studies are being performed in Europe and Asia. Included are the following: 45 "immunotherapy/cytokine" protocols, most of which involve the transfer of cytokine genes into tumor cells; ten "tumor suppressor (anti-oncogene)/antisense" protocols; five "drug resistance" protocols, all of which involve transfer of multidrug resistance cDNA; and 26 "gene marker" protocols primarily designed to assess the efficacy and safety of gene transfer and gene therapy in patients. All these protocols involve transfer of the neomycin resistance gene into lymphocytes or stem cells ex vivo and their readministration to patients. Questions related to the distribution and persistence of the genetically altered cells, identified by their resistance to neomycin toxicity, are being addressed. There are also 15 "drug sensitivity" protocols; all 15 involve retroviral transfer of the herpes simplex virus (HSV) thymidine kinase gene (HSV1-tk) into tumor cells followed by systemic treatment with the prodrug ganciclovir. Their use in cancer therapy is to create significant differences between normal and malignant cells by selective transduction of the "susceptibility" gene into malignant, but not normal cells.

11.2 Imaging the Expression of Transduced Genes

A noninvasive, clinically applicable method for imaging the expression of successful gene transduction in target tissue or specific organs of the body would be of considerable value; it would facilitate the monitoring and evaluation of gene therapy in human subjects by defining the *location*, *magnitude*, and *persistence* of gene expression over time. Targeting gene therapy to particular tissue (e.g., tumor) or specific organs is an increasingly active area of research, with over 400 related articles published in 1994 and over 800 articles in 1995 based on a MEDLINE search. Successful targeting of cells in tissue culture can be assayed in several different ways, including direct assays for the gene product or indirect assays for mRNA (northern blot analysis) and functional assays (e.g., drug sensitivity). These assays require tissue culture techniques or sampled tissue and therefore have limited application in a clinical setting. They also provide no spatial information with respect to the location of successful gene transfection in specific tissue or organs. Several histochemical techniques have been developed to identify the location of transduced cells and gene expression in tissue sections obtained in animals. The most popular "marker genes" for histochemical assays in tissue sections are the *Escherichia coli lacZ* gene (Price et al. 1987) and the firefly luciferase gene (Manthorpe et al. 1993).

11.3 Potential Clinical Impact

Several issues that are important for clinical optimization of gene therapy remain unresolved in many current clinical protocols:

1. Has gene transduction or transfection been successful?
2. Is the distribution of the transduced or transfected gene localized to the target organ or to target tissue, and is the distribution in the target optimal?
3. Is the level of gene expression in the target organ or tissue sufficient to result in a therapeutic effect?
4. In the case of combined prodrug–gene therapy protocols, when is gene expression maximum (optimal) and when is the optimal time to initiate treatment with the prodrug?
5. How long does gene expression persist in the target and other tissues?

Currently there are *no* proposals in the literature that adequately address the important clinical issues listed above. We propose that noninvasive imaging techniques using selected "marker gene" and "marker substrate" combinations will provide a practical and clinically useful way to identify successful gene transduction and expression in patients undergoing gene therapy.

To illustrate the principles involved in imaging the expression of foreign genes, we will focus on the HSV1-tk–ganciclovir "drug sensitivity" cancer treatment protocol. This is an ideal gene therapy protocol and can serve as a model system for monitoring gene expression by noninvasive imaging.There are 15 approved protocols involving retroviral transfer of the HSV1-tk gene into tumors of patients(Clinical Protocol List 1995). Eight of these involve the treatment of brain tumors, while the other protocols are for the treatment of leptomeningeal carcinomatosis, persistent head and neck cancer, advanced mesothelioma, and ovarian cancer. This approach is based on the fact that certain genes can be used to "sensitize" cells to drugs that are normally inactive or nontoxic (e.g., prodrug), and these genes have been described as "susceptibility" or "suicide" genes. Most susceptibility genes encode viral or bacterial enzymes that convert inactive forms of a drug (prodrug) into toxic compounds or antimetabolites capable of inhibiting nucleic acid synthesis. HSV thymidine kinase (HSV-TK) selectively converts the normally nontoxic prodrug ganciclovir into toxic compounds that results in cell death. These so-called drug sensitivity or susceptibility gene therapy protocols depend on the expression of HSV-tk in the target (tumor) tissue.

Three issues that are important for clinical optimization of HSV-tk gene therapy remain unresolved:

1. Has retroviral HSV-tk gene transduction been successful?
2. Is the level of HSV-tk expression adequate to achieve the desired therapeutic effect with ganciclovir?
3. When is the optimal time to initiate treatment with ganciclovir (following transduction or transfection when is HSV-tk most widely expressed)?

There are no noninvasive methods currently available to address these questions. The ability to selectively image HSV-tk expression could address these questions, and the imaging results could have a significant impact on patient management and could alter patient outcomes.

11.4 Characteristics of a Good "Marker Gene" and "Marker Substrate"

Clinical monitoring of gene expression requires the appropriate *combination* of "marker gene" and "marker substrate." We consider the following characteristics to be ideal, if not essential (Tjuvajev et al. 1995):

1. A marker gene is usually a "foreign gene" that is not present (or not normally expressed) in the host tissue, and the "gene product" is an enzyme that can be expressed in transduced or transfected host cells. This enzyme must be nontoxic to host cells and catalyze a reaction with the marker substrate where the reaction product accumulates within transduced or transfected cells.
2. A marker substrate is chosen to match the marker gene; it is a compound that is not metabolized (or only slowly metabolized) by the host and does not accumulate in nontransduced or nontransfected host tissue. It can be radiolabeled with appropriate isotopes for

clinical imaging using gamma camera, single photon emission computed tomography (SPECT), or photon emission tomography (PET) techniques, or is a paramagnetic compound appropriate for clinical magnetic resonance imaging (MRI).

3. The marker substrate must cross cell membranes readily, be rapidly metabolized by the marker gene product, and be effectively trapped within transduced cells throughout the period of imaging; in addition, it must accumulate to levels that are measurable by existing clinical imaging techniques.
4. The accumulation of the marker substrate in transduced or transfected cells must reflect the activity of the gene product and hence the expression of the marker gene in transduced or transfected tissue.

11.5 Selection of HSV1-tk as the Marker Gene

To demonstrate the feasibility of imaging gene expression, we initially established model systems in tissue culture and in experimental animals (Tjuvajev et al. 1995). HSV1-tk was selected as an example of a marker gene, and we recently evaluated three potential marker substrates for the enzyme HSV1-TK. HSV1-tk has been extensively studied; it is nontoxic in humans and is currently being used as a susceptibility gene (in combinationwith ganciclovir) in nine clinical gene therapy protocols (see above). HSV1-tk can be used as a marker gene as well as a therapeutic gene (Moolten 1986; Borrelli et al. 1988; Moolten and Wells 1990; Culver et al. 1992). In gene therapy protocols using HSV1-tk as a susceptibility gene, identifying the location and magnitude of HSV1-TK expression by noninvasive imaging would be highly desirable and would provide a basis from which the timing of ganciclovir treatment can be optimized. This represents an ideal situation in which the therapeutic and marker genes are the same.

Since all therapeutic genes may not have an appropriate marker substrate that can be labeled for imaging (as we have shown for HSV1-tk), we are exploring an alternative approach by using a double gene construct in which the therapeutic gene is linked to a second marker gene by an internal ribosomal entry site (IRES) sequence (Ghattas et al. 1991; Sokolic et al. 1996; Sugimoto et al. 1995a). It is important to note that vectors containing double gene constructs have been used in the past to provide a "selection" gene, e.g., *NeoR* or *MDR1*, to facilitate isolation of cells (Sugimoto et al. 1995a; Mulligen and Berg 1981) or a susceptibility or suicide gene [e.g., *HSV1-tk* (Mullen et al. 1992) or *cytosine deaminase* (Haberkorn et al. 1996) to provide a safety mechanism for eliminating transduced cells and a way to control unwanted retroviral spread], along with a therapeutic or other gene of interest (Sugimoto et al. 1995b). In this context, HSV1-tk has been used as a suicide gene.

11.6 Choice of Marker Substrates for Imaging HSV-TK Expression

The 2'-fluoro-1-β-D-arabinofuranosyl and 2'-fluoro-1-β-D-ribofuranosyl analogues of uridine were selected for study because they were developed as potential antiviral agents, and some have been radiolabeled and studied as potential imaging agents of viral infections in the past (Iwashina et al. 1988; Saito et al. 1982, 1984; Price et al. 1983; Tovell et al. 1987). They are selectively phosphorylated by HSV-TK and poorly or not at all by mammalian thymidine kinase, and they accumulate well in virally infected cells and minimally in noninfected dividing cells. They can be radiolabeled with several different radionuclides, including ^{11}C and ^{18}F, or carrier-free ^{131}I, ^{123}I, and ^{124}I, appropriate for clinical imaging with PET (or SPECT), which can be performed in many nuclear medicine departments. The longer half-life of the iodine radionuclides provides an advantage with respect to production and distribution of the radiopharmaceutical, and with respect to the clearance of radiolabeled metabolites prior to imaging. An important characteristic of the 2'-fluoro substitution on the arabinose or ribose is that it provides a protective group against enzymatic cleavage of the N-glucosidic bond by nucleoside phosphorylases. The increased stability of the N-glucosidic bond results in a significant prolongation of plasma half-life of these compounds, and they are excreted largely unchanged in the urine (Table 1). The plasma clearance data clearly indicate that a substantially higher input function (blood concentration-time integral) is obtained with 5-iodo-2'-fluoro-2'-deoxy-1-β-D-arabinofuranosyl-uracil (FIAU), 2'-fluoro-5-iodovinyl-1-β-D-ribofuranosyl-uracil (FIVRU), [^{11}C]-5-

Table 1. Comparison of nucleoside clearance from plasma in rodents and humans

Nucleoside	Plasma half-time (min)	24-hour excretion of radioactivity (%) (urine)[a]	Reference
TdR[b]	5–7	–	Molnar et al. 1984
IUdR[b]	5–10	~70[c]	Tjuvajev et al. 1993
FMAU[b]	260	~85[d]	Saito et al. 1984
FEAU[b]	78	~33[d]	Kong et al. 1992
FIAU[e]	168	~73[d]	Feinberg et al. 1985
FIVRU[f]	89	44 (0–5h)[d]	Iwashina et al. 1988
Ganciclovir[g]	210	91[d]	Fletcher et al. 1986
Acyclovir[g]	150	62–91[h]	Blum et al. 1982

TdR, thymidine; IUdR, 5-Iododeoxyuridine; FMAU, 2'-fluoro-5-methyl-1-β-D-arabinofuranosyl-uracil; FEAU, 2'-fluoro-5-(2-fluoro-ethyl)-1-β-D-arabinofuranosyl-uracil); FIAU, 2'-fluoro-5-iodo-1-β-D-arabinofuranosyl-uracil); FIVRU, 2'-fluoro-5-iodovinyl-1-β-D-ribofuranosyl-uracil; ganciclovir, 9-(1,3-dihydroxy-2-propoxymethyl)-guanine; acyclovir, 1-Amino-1,9-dihydro-9-[(2-hydroxyethoxy)methyl]-6H-purine-6-one

[a]% of administered dose.

[b]Values based on data obtained in rats.

[c]Predominantly as iodide.

[d]Predominantly as parent compound.

[e]Calculated values based on FIAC studies in humans; FIAC is rapidly deaminated to FIAU in humans.

[f]Calculated values based on data in mice.

[g]Values based on data obtained in humans.

[h]Up to 14% may be due to 9-carboxymethoxymethylguanine.

methyl-1-β-D-arabinofuranosyl-uracil (FMAU), and [^{11}C]-5-ethyl-1-β-D-arabinofuranosyl-uracil (FEAU) in comparison to 5-iodouracil-2'-deoxyriboside (IUdR) and thymidine. As a result, a greater amount of nondegraded radiolabeled drug is delivered to the target tissues, and confounding problems associated with imaging radiolabeled metabolites in both target and surrounding tissue are avoided.

Our preliminary results demonstrate the limitation of ^{11}C-labeled 2'-fluoro analogues, e.g., FMAU or FEAU, due to the short half-life (20 min) of [^{11}C] and the advantage for "later imaging," after a period of time to allow for body clearance of the nonphosphorylated analogue (and any radiolabeled metabolites that may have formed). Images 1 h after i.v. administration are likely to reflect substantial amounts of nonphosphorylated and unincorporated [^{11}C] radioactivity with FMAU and FEAU, despite the favorable sensitivity and specificity indices for FEAU that were measured in vitro (see Fig. 3 and Table 2). The question of whether 4-h [^{18}F] images, after 2'-fluoro-5-fluoro-1-β-D-arabino-furanosyl-uracil (FFAU) or 2'-fluoro-5-(2-fluoro-ethyl)-1-β-D-arabino-furanosyl-uracil (FFEAU) administration, will provide sufficient specificity for HSV1-tk expression remains to be more fully assessed. One reason for pursuing studies with [^{18}F]FFAU and [^{18}F]FFEAU is the wider availability and substantially lower cost of ^{18}F labeling compared to ^{124}I labeling of 2'-fluoro nucleosides for PET imaging. Very late imaging 12–48 h after i.v. administration following an extended period of "washout" *cannot* be performed with either [^{11}C]- or [^{18}F]-labeled 2'-fluoro nucleosides.

11.7 Other "Marker Substrates" for Imaging HSV-TK Expression

We considered other substrates that might be selectively phosphorylated by the enzyme HSV1-TK, including a number of thymidine and guanosine analogues. We evaluated two non-2'-fluoro substrates (iododeoxyuridine and ganciclovir) in preliminary studies (see Fig. 3, Table 2; Tjuvajev et al. 1995). IUdR showed good sensitivity, but no selectivity since it is phosphorylated by wild-type cells as readily as by HSV1-tk-transduced cells. Ganciclovir showed modest selectivity, but very low sensitivity compared to FIAU. Similar results with ganciclovir have been reported by others (Haberkorn et al. 1996). Other guanosine analogues have been labeled and studied, including 8-[^{18}F]acyclovir (Srinivasan et al. 1996) and 9-[(3-[^{18}F]fluoro-1-hydroxy-2-propoxy)-methyl]-guanine (Alauddin et al. 1996), as well as 5-fluoro-6-[^{3}H]-2'-deoxy-cytidine (Haberkorn et al. 1996). In our initial review and consideration of potential marker substrates for measuring/imaging HSV1-TK activity (including the above-mentioned compounds), we focused on several factors important for both sensitivity and selectivity. The sensitivity criteria included clearance measurements

[K_i, ml/min per g cells (or tissue)] of the radiolabeled marker substrate from the incubation medium (in vitro studies) or plasma (in vivo studies), the biological clearance (half-life) of the marker substrate and radiolabeled metabolites from plasma, the physical half-life of the radionuclide used for labeling, the time of measurement or imaging, and the contribution of background (nonphosphorylated or non-DNA-incorporated radioactivity) to the measured signal. Selectivity criteria were established by comparing HSV1-tk-transduced cells or tumor tissue with corresponding nontransduced cells or tumor tissue. A model was developed which could compare the effects of these different factors on the level of radioactivity that would be expected in both transduced and nontransduced tissue at the time of imaging. Although a complete data set is not available (at this time) for a direct comparison between 2'-fluoro nucleoside and guanosine analogues, the literature (Haberkorn et al. 1996; Srinivasan et al. 1996; Alauddin et al. 1996) and our preliminary results (see Figs. 3, 4), the existing data suggest that HSV1-TK phosphorylation of the 2'-fluoro nucleosides is more rapid than the guanosine analogues acyclovir and ganciclovir and that a higher fraction of 2'-fluoro nucleosides are incorporated into DNA due to the chain termination effects of acyclovir and ganciclovir incorporation. Using [^{18}F] to label the marker substrate and imaging at 2 or 4 h after i.v. injection was shown to result in lower sensitivity and selectivity due to higher tissue background, compared to labeling the same substrate with [*I] and imaging at 24 h. This difference was particularly noticeable when the plasma clearance half-time of the radiolabeled substrate and/or radiolabeled metabolites was in the range of 2–4 h (see Table 1) and may be a potential problem with [18F]-labeled analogues of ganciclovir and acyclovir. For these reasons, the 2'-fluoro nucleoside analogues of uracil were selected for study.

11.8 In Vitro Experimental Studies

11.8.1 Reagents

The RG2 glioma cell line was originally derived from an ethylnitrosourea (ENU)-induced rat glioma (Ko et al. 1980) and was kindly supplied by Dr. Darell Bigner (Duke University Medical Center, Durham, NC). The gp-STK-A2 producer cell line, which produces the recombinant STK retrovirus containing HSV1-tk, was kindly supplied by Dr. Frederick Moolten (Memorial Veterans Hospital, Bedford, MA). The STK retrovirus also contains a *NeoR* gene (which encodes resistance to the neomycin analogue geneticin disulfate) and the SV40 early region promoter and enhancer (Moolten and Wells 1990). The transduction of the HSV1-tk gene into wild-type RG2 cells, cell culture procedures for creating the bulk culture RG2TK$^+$ cell line, and the isolation of HSV1-TK-expressing clones has recently been described (Tjuvajev et al. 1995). Similar procedures were used to transduce W256 cells and establish the W256TK$^+$ bulk culture cell line.

11.8.2 Independent Assays of HSV1-tk Gene Expression in Different RG2TK Cell Lines

Two independent assays for HSV1-tk gene expression were performed in several different RG2TK cell lines: (a) sensitivity to the antiviral drug ganciclovir and (b) measurements of HSV-TK mRNA. Ganciclovir sensitivity is a functional assay for the presence of HSV1-TK (cells that do not express HSV1-TK have a very high ED_{50}). HSV-TK mRNA expression is a specific, but indirect measure of HSV1-TK. Both measures were shown to be highly correlated with marker substrate accumulation in different RG2TK cell lines in culture (Tjuvajev et al. 1995). Similarly, both measures could be correlated with marker substrate accumulation in i.c. and s.c. tumors in animals, tumors produced by the same cell lines studied in tissue culture (see Table 3 and Fig. 6).

11.8.3 Uptake of Marker Substrates for Identifying HSV1-tk Gene Expression in Tissue Culture

Three potential marker substrates (FIAU, IUdR, and ganciclovir) for the enzyme HSV1-TK were previously evaluated, and the methods used in these studies (and proposed in this application) described (Tjuvajev et al. 1995). An example of one triplicate experiment measuring the time course of FIAU and thymidine accumulation, expressed as a cell-to-medium radioactivity ratio, in nontransformed RG2 cells and transformed

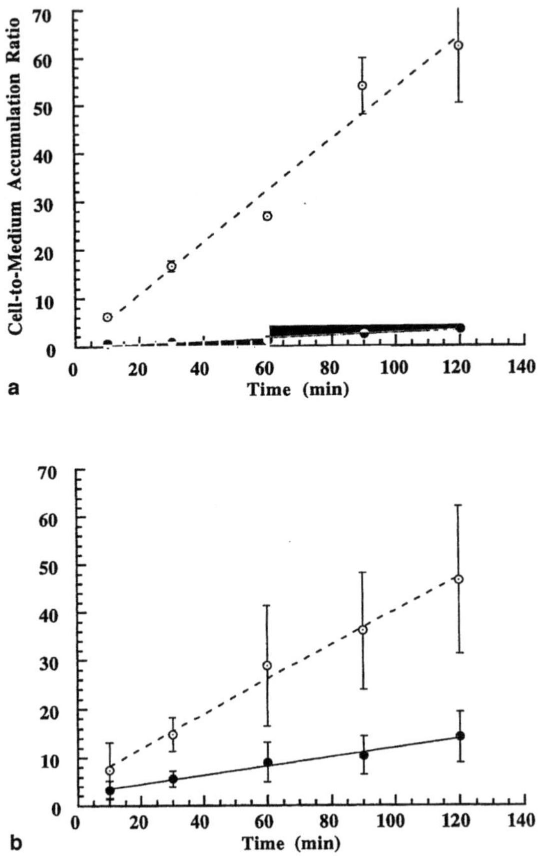

Fig. 1a,b. Cell-to-medium ratio of nucleoside accumulation in **a** nontransformed RG2 cells and **b** transformed RG2TK+ cells as a function of time for 5-iodo-2'-fluoro-2'-deoxy-1-β-D-arabinofuranosyl-uracil (FIAU, *closed symbols*) and thymidine (*open symbols*)

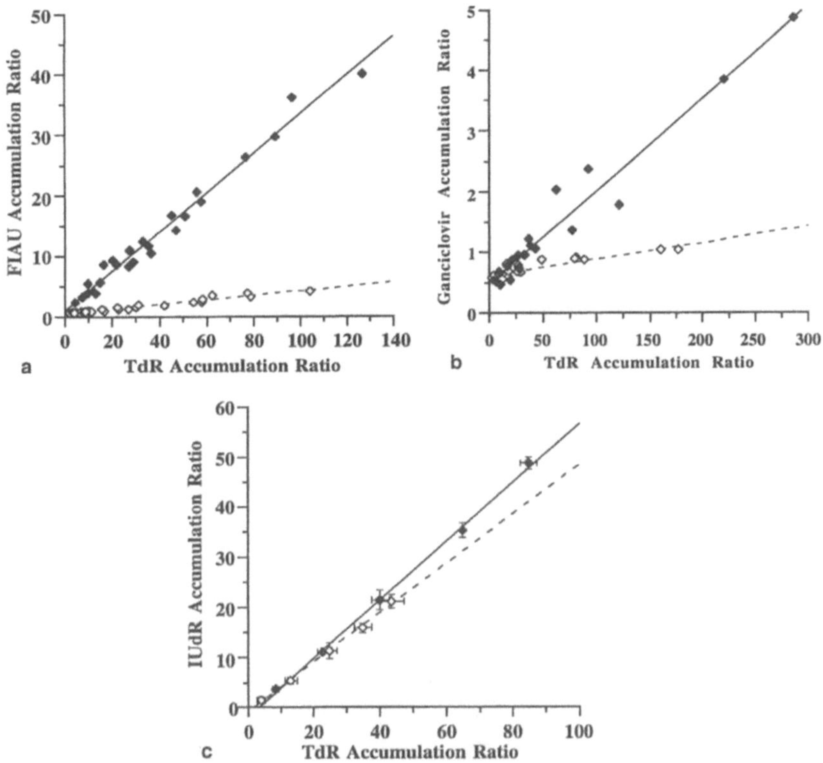

Fig. 2a–c. Comparison between the accumulation ratio of **a** 5-iodo-2'-fluoro-2'-deoxy-1-β-D-arabinofuranosyl-uracil (FIAU), **b** ganciclovir, and **c** 5-iodouracil-2'-deoxyriboside in transformed RG2TK+ (*closed symbols*) and non-transformed RG2 cells (*open symbols*). *TdR*, thymidine

RG2TK+ cells is shown in Fig. 1. The slope of the plots represents the medium-to-cell clearance rate (ml medium/min per g cells).

It should be noted that cell proliferation varied somewhat between experiments for both transduced RG2TK+ and nontransduced RG2 cell cultures. Since cell proliferation assayed by cell counting techniques correlated with thymidine incorporation (data not shown), we normal-

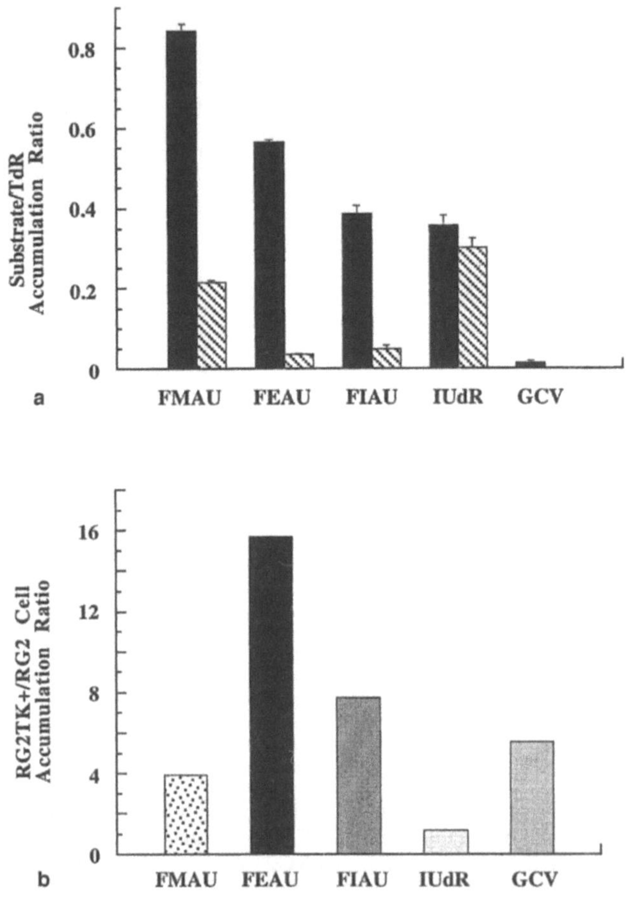

Fig. 3. a Substrate to thymidine (*TdR*) ratio for different compounds in transformed RG2TK (*black bars*) and nontransformed RG2 cells (*shaded bars*). **b** RG2TK+/RG2 cell accumulation ratio. *FMAU*, 5-methyl-1-β-D-arabinofuranosyl-uracil; *FEAU*, 5-ethyl-1-β-D-arabinofuranosyl-uracil; *FIAU*, 5-iodo-2'-fluoro-2'-deoxy-1-β-D-arabinofuranosyl-uracil; *IUdR*, 5-iodouracil-2'-deoxyriboside; *GCV*, ganciclovir

ized nucleoside analogue accumulation to that of thymidine in order to account for differences in cell proliferation between experiments. A comparison between the accumulation of different nucleoside analogues in $RG2TK^+$ transduced cells and nontransduced RG2 cells is shown in Fig. 2. FIAU accumulation in bulk $RG2TK^+$ cells was 7.8 times greater than that in wild-type RG2 cells (Fig. 2a). Ganciclovir accumulation was 5.6-fold greater in $RG2TK^+$ cells than in RG2 cells (Fig. 2b), whereas the difference was only 1.2-fold for IUdR (Fig. 2c). The accumulation of ganciclovir by $RG2TK^+$ cells was *small* compared to that of FIAU and IUdR (less than 5% and 3%, respectively) and may *not* provide a sufficient signal for imaging. Although the accumulation of IUdR by $RG2TK^+$ cells was somewhat higher than that of FIAU in this system, the accumulation of IUdR does not distinguish between transduced $RG2TK^+$ and nontransduced RG2 cells. This suggests that IUdR is not an effective marker substrate for identifying selective HSV-TK gene expression in this system.

We obtained preliminary data on two additional potential marker substrates, FMAU and FEAU, because these radiolabeled 2'-fluoro nucleoside analogues (methyl-[^{3}H]FMAU and 2-[^{14}C]FEAU) were synthesized and recently became available. The results of our studies in HSV-tk-transduced RG2 ($RG2TK^+$) cells and wild-type, nontransduced RG2 cells is shown in Fig. 3. The substrate accumulation ratio, normalized for thymidine accumulation to account for differences in cell proliferation rates in the culture plates, is shown for $RG2TK^+$ cells and RG2 cells, respectively (Fig. 3a). Substrate accumulation in the two cell lines is compared by the $RG2TK^+$/RG2 cell accumulation ratio (Fig. 3b). Both FMAU and FEAU accumulate to higher levels in transduced cells than FIAU or the other substrates (Fig. 3a). However, FMAU is also accumulated to substantial levels by nontransduced RG2 wild-type cells, levels that are considerably higher than those of FEAU and FIAU. The $RG2TK^+$/RG2 "selectivity ratio" is highest for FEAU followed by FIAU, whereas the $RG2TK^+$/RG2 ratio for FMAU is considerably lower (Fig. 3b). Ganciclovir has very low accumulation (Fig. 3a) and only modest "selectivity" (Fig. 3b) compared to FEAU and FIAU.

Table 2. Substrate senstivity and selectivity for HSV1-TK

Substrate	Sensitivity[a,b]	Selectivity[a,c]
FIAU	0.0090±0.0003	0.238±0.027
Ganciclovir	0.0004±0.00003*	0.143±0.054
IUdR	0.0029±0.0034*	0.006±0.009*
FMAU[d]	0.0188±0.0013*	0.087±0.002*
FEAU[d]	0.0161±0.0008*	0.447±0.037*

FIAU, 5-iodo-2'-fluoro-2'-deoxy-1-β-D-arabinofuranosyl-uracil; IUdR, 5-iodouracil-2'-deoxyriboside; FMAU, [^{11}C]-5-methyl-1-β-D-arabinofuranosyl-uracil; FEAU, [^{11}C]-5-ethyl-1-β-D-arabinofuranosyl-uracil.
[a]Data from Fig. 3.
[b]Δ (substrate/thymidine) / Δ (HSV1-TK mRNA/S28 RNA IOD).
[c]Sensitivity / (substrate/TdR) for RG2.
[d]New preliminary data, not previously presented.
*Significantly different from FIAU values (*p*.05).

11.8.4 Data Analysis

In order to evaluate the proficiency of marker substrates for imaging of HSV1-tk expression, two parameters were developed:

1. *Sensitivity*, defined as the change in substrate uptake (normalized to thymidine) divided by the change in HSV1-tk expression
2. *Selectivity*, defined as sensitivity divided by substrate uptake due to endogenous (mammalian) TK

These two related measures were considered useful for imaging considerations because both intensity (i.e., high sensitivity) and contrast (the ability to differentiate between HSV1-TK and endogenous TK, i.e., high selectivity) in the images are required. It is important to note that the selectivity measure should not be interpreted as a relative sensitivity of the substrate to HSV1-TK versus endogenous TK. Also, neither of the measures represents a fraction of some ideal value (i.e., they do not have a maximum of one). They are intended for the direct comparison of substrates only.

A summary comparison of the sensitivity and selectivity is shown in Table 2. Note that the change in HSV1-TK expression was taken to be

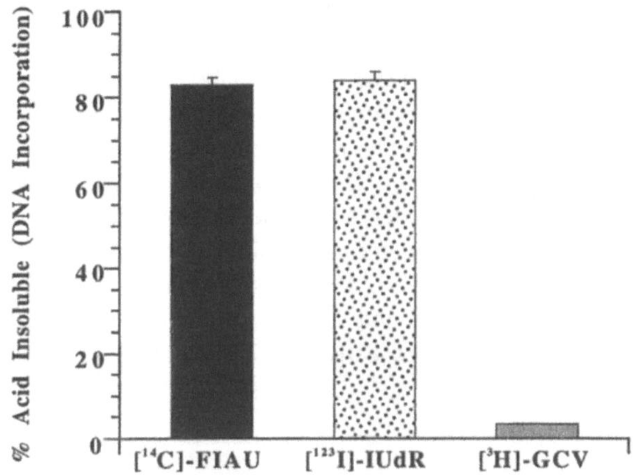

Fig. 4. Substrate incorporation into the acid insoluble cell fraction of RG2TK+ cells. *FIAU*, 5-iodo-2'-fluoro-2'-deoxy-1-β-D-arabinofuanosyl-uracil; *IUdR*, 5-iodouracil-2'-deoxyriboside; *GCV*, ganciclovir

the difference in the HSV1-TK mRNA/S28 RNA integrated optical density (IOD) ratios. From these results, it can readily be seen that (a) the accumulation of ganciclovir by $RG2TK^+$ cells was small compared with the other substrates and may not provide a sufficient signal for imaging and (b) IUdR and FMAU do not distinguish well between transduced $RG2TK^+$ and nontransduced RG2 cells.

11.8.5 Phosphorylated Metabolites and Incorporation of Radioactivity into DNA

In a triple-label, 120-min in vitro experiment, the accumulation of 2-[^{14}C]FIAU was directly compared with that of [^{123}I]IUdR and 8-[^{3}H]ganciclovir in $RG2TK^+$ and RG2 cells. The acid-insoluble fractions of FIAU and IUdR accumulation in both cell lines at 2 h were similar, ranging between 80% and 90% of total activity, whereas the acid-insoluble fraction after incubation with ganciclovir was much lower (Fig. 4; data for $RG2TK^+$ cells is shown). This suggests that most of the measured FIAU- and IUdR-derived radioactivity is in DNA, in contrast to that derived from ganciclovir. It has also been recently reported by others (Haberkorn et al. 1996) that only a small fraction of cell-derived radioactivity is in the acid-insoluble fraction when transduced, HSV1-TK-expressing Morris hepatoma cells are incubatedwith 8-[^{3}H]ganciclovir. High-performance liquid chromatography (HPLC) analysis of the soluble fraction demonstrated a time-dependent shift of ganciclovir radioactivity to its phosphorylated metabolites in transduced cells (but not in wild-type control cells); the phosphorylated metabolites accounted for 3.6%, 14%, 42%, and 91% of soluble radioactivity after 2, 4, 24, and 48 h incubation, respectively (U. Haberkorn, personal communication).

11.8.6 Relationship Between FIAU Uptake and Two Independent Measures of HSV1-tk Expression

On the basis of the sensitivity and selectivity indices obtained in RG2TK^{+} cells (Table 2) and the availability of 2-[^{14}C]FIAU from Moravek Biochemicals Inc. at the time these experiments were performed, we tested whether there was a consistent relationship between the level of HSV1-tk expression in different cell lines and the level of FIAU accumulation in these cell lines. Several RG2TK clones, as well as wild-type RG2 and RG2TK^{+} cells, were compared with respect to FIAU accumulation and two independent measures of HSV-tk expression, ganciclovir sensitivity (ED_{50}) and normalized IOD values of HSV1-TK mRNA in corresponding cell lines (Fig. 5). Accumulation of FIAU, expressed as the FIAU/thymidine uptake ratio (slope in Fig. 3), was proportional to the levels of HSV1-TK mRNA in corresponding cell lines (Fig. 5a) and could be described by the equation:

$$\text{FIAU/thymidine uptake ratio} = 0.0449 + 0.00861 * \text{IODtk}$$

As expected, a highly significant inverse relationship was also observed between FIAU uptake and ganciclovir sensitivity (Fig. 5b) and could also be described by the equation:

$$y = 0.325 – 0.121 \log(x)$$

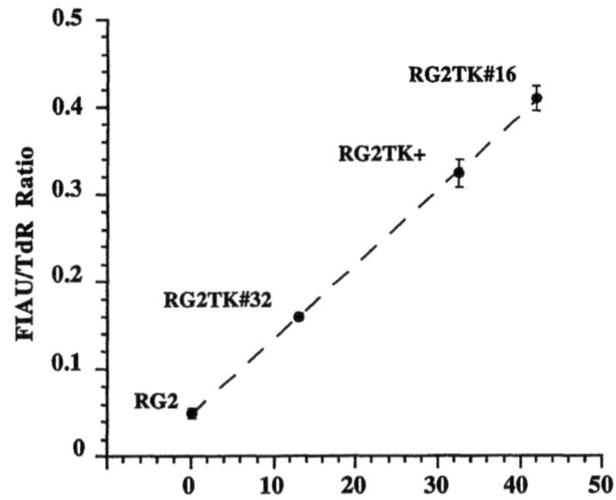

a

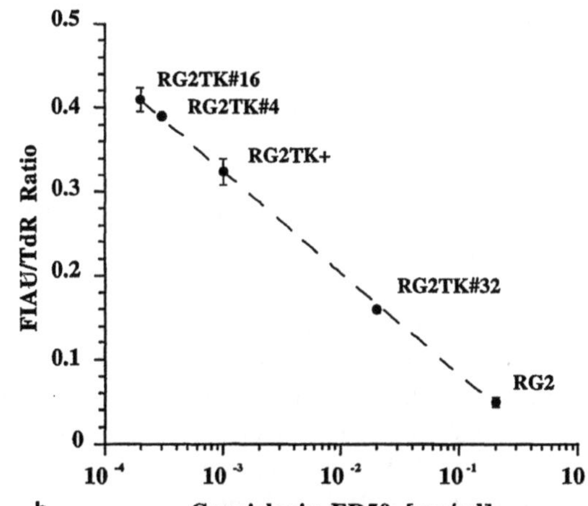

b

Fig. 5a,b. Comparison of 5-iodo-2'-fluoro-2'-deoxy-1-β-D-arabinofuanosyl-uracil *(FIAU)* accumulation with **a** HSVi-TK mRNA and **b** sensitivity to ganciclovir. *IUdR*, 5-iodouracil-2'-deoxyriboside; *TdR*, thymidine

These results indicate that the level of FIAU accumulation in different transduced cell lines correlates with the level of HSV1-tk expression as measured by HSV1-TK mRNA on northern blot analysis and by sensitivity to ganciclovir, a functional measure of HSV1-TK expression, in corresponding cell lines.

11.9 In Vivo Experimental Studies

11.9.1 Animal Preparation and Tumor Inoculation

Male Fisher 344 rats or immunocompromised R-Nu rats were anethetized with a gas mixture (2% isoflurane, 68% nitrousoxide, and 30% oxygen), and a tumor cell suspension (10^5 cells in 10 μl) was slowly injected over 2 min into the caudate nucleus under stereotactic control, as previously described (Tjuvajev et al. 1993). Subcutaneous inoculations (10^6 cells in 50 μl) were also performed. Following the procedure, the animals were marked, placed in standard cages with water and Rodent Lab Chow #5001 (Purina Mills, Inc.), and observed daily.

11.9.2 Selective Imaging of HSV1-TK Expression with FIAU in Experimental Animals

On the basis of the sensitivity and selectivity values obtained in vitro (see Table 3) and the availability of 2-[^{14}C]FIAU at the time these experiments were performed, all in vivo studies were performed with FIAU as the marker substrate for HSV1-tk expression. Fisher 344 rats were prepared with i.c. RG2TK$^+$ and RG2 tumors as described above. Sixteen days after RG2TK$^+$ and RG2 cell inoculation, each animal was injected i.v. with 50 μCi 2-[^{14}C]FIAU and killed with euthanasia solution 24 h later. The brain was rapidly extracted and frozen in preparation for cryosectioning and autoradiography (Blasberg et al. 1980, 1981). The dried tissue sections were placed in an X-ray cassette along with 15 ^{14}C-methyl methacrylate standards of known, previously calibrated radioactivity (4.4–2354 nCi/g tissue) and exposed to SB5 single-coated X-ray film (Kodak) for 10 days. The resultant autoradiographic images reflect the spatial distribution of FIAU-derived ^{14}C radioactivity in the

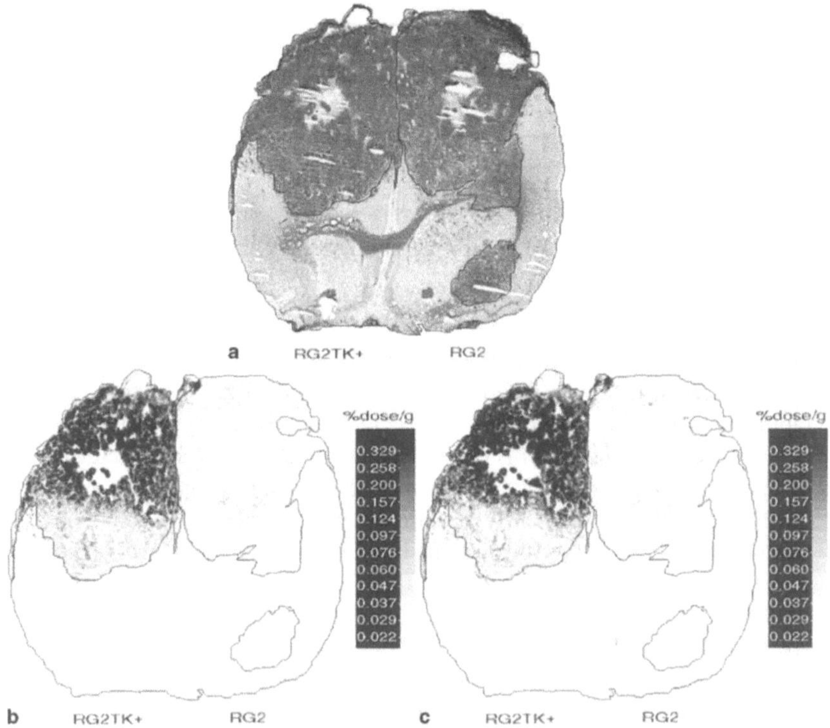

Fig. 6a–c. Imaging HSV1-tk gene expresssion with 2-[^{14}C]5-iodo-2'fluoro-2'-deoxy-1-β-D-arabinofuanosyl-uracil (FIAU). **a** Histology. **b** FIAU autoradiography, 24-h study, nonwashed. **c** 2-[^{14}C]FIAU autoradiography, 24-h study, washed for 4 h

tissue sections. To convert the X-ray images to tissue radioactivity (nCi/g), the optical density of the standards was measured by computerized image analysis. A standard curve that relates mean optical density and tissue radioactivity was generated for each film. Knowing tissue radioactivity and the injected dose of 2-[^{14}C]FIAU, the autoradiographic images are readily converted to parametric images of percentage dose per gram tissue and color-coded to a range of values.

The ability to selectively image the expression of a marker gene, HSV1-tk, in vivo is shown in Fig. 6. The histology and autoradiographic

Table 3. FIAU senstivity and selectivity for HSV1-TK expression in vivo

Substrate	RG2TK+/RG2 i.c. tumors[a]		RG2TK#16/RG2 i.c. tumors[b]	
	Sensitivity[c]	Selectivity[d]	Sensitivity[e]	Selectivity[d]
FIAU	0.0039	0.165	0.0058	0.222

FIAU, 5-iodo-2'-fluoro-2'-deoxy-1-β-D-arabinofuranosyl-uracil.
[a]See Fig. 6.
[b]Figure not shown.
[c]Δ (substrate % dose/g) / Δ (HSV1-TK mRNA/S28 RNA IOD).
[d]Sensitivity / uptake in RG2 tumor.

images were registered so that all regions of interest (ROIs) drawn on the histologic image would outline equivalent regions on the autoradiographic images. The outline of the RG2TK$^+$ and RG2 tumors in each hemisphere of the rat brain is clearly seen in the toluidine blue-stained histological section (Fig. 6a). In the autoradiographic images (Fig. 6b), the RG2TK$^+$ tumor is clearly visualized, whereas the RG2 tumor is barely detectable and the surrounding brain is at background levels. The punctate heterogeneous levels of FIAU activity observed in the RG2TK$^+$ tumor (excluding the white areas representing tissue necrosis) are thought to reflect differences in HSV1-tk gene expression in the nonclonal, mixed population of transduced RG2TK$^+$ cells growing intracerebrally. Quantitative analysis yielded mean activities of 0.180±0.140 and 0.022±0.024% dose/g for the RG2TK$^+$ and RG2 tumors, respectively. Peak activity in the RG2TK$^+$ tumor was 0.43% dose/g; adjacent brain activity was below the level of detection (.005% dose/g) on the autoradiograms.

The tissue section used for autoradiography in Fig. 6c was adjacent to the one used in Fig. 6b; it was rinsed in acid solution (10% tricarboxylic acid, TCA) for 4 h to remove nonincorporated FIAU and radiolabeled metabolites prior to autoradiographic exposure. Rinsing tissue sections had little effect on the distribution and amount of radioactivity measured in the autoradiograms (compare Figs. 6b and 6c). This indicates a very low background activity due to soluble radiolabeled metabolites and demonstrates the value of using an "in vivo washout" strategy (imaging 24 h after FIAU administration) to reduce background radioactivity.

The sensitivity and selectivity measures for the autoradiographic data are shown in Table 3. These values are in the same units as the tissue sampling results (Table 2) and are based on measurements of percentage dose per gram tissue in three animals (two $RG2TK^+$ and one RG2TK#16).

11.9.3 Imaging HSV1-tk Expression Following Transduction of Preestablished Wild-Type RG2 Tumors

To assess whether FIAU imaging was sufficiently sensitive to monitor expression of the HSV1-tk gene following in vivo transduction of preestablished RG2 tumors with the STK retrovirus, three different experiments were performed. The *first series* involved the coimplantation of wild-type RG2 cells and gp-STK-A2 vector-producer cells (1:1 ratio) into the right hemisphere. The left hemisphere was inoculated with gp-STK-A2 vector-producer cells only. Ten days after inoculation, animals were injected i.v. with 50 μCi 2-[^{14}C]FIAU and killed 24 h later, and the brain was processed for quantitative autoradiography (QAR) and histology. The HSV1-tk-transfected tumor was well visualized on the autoradiographic image (Fig. 7b). In this case, the pattern of FIAU activity reflects spatial heterogeneity in the efficiency of retroviral transfection as well as differences in HSV1-tk gene expression. Surrounding normal brain and the area inoculated only with gp-STK-A2 vector-producer cells are poorly or not at all visualized. Histologic analysis of this area (see Fig. 10a) showed an inflammatory reaction with perivascular infiltrates; the outlined area was heavily infiltrated, and almost all vector-producer cells were rejected. Rinsing tissue sections in acid solution for 4 h prior to X-ray film exposure had little effect on the distribution and amount of radioactivity measured in the transduced tumor, but eliminated the low level of radioactivity localized to the area of the left hemisphere that was inoculated with gp-STK-A2 vector-producer cells (Fig. 7c), indicating that the low-level activity in this region of the left hemisphere was not incorporated into macromolecules. Quantitative analysis yielded mean activities of 0.34±0.24% dose/g for the in vivo transduced RG2(TK^+) and 0.028±0.023% dose/g for the gp-STK-A2 vector-producer cell-inoculated area prior to rinsing. Peak activity in the

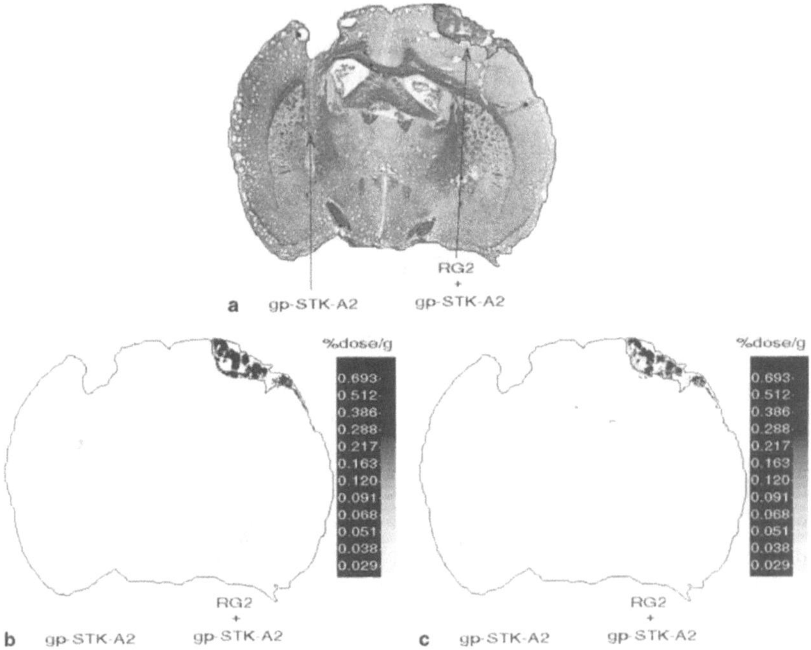

Fig. 7a–c. Imaging HSV1-tk gene transduction and expresssion with 2-[^{14}C]5-iodo-2'fluoro-2'-deoxy-1-β-D-arabinofuanosyl-uracil (FIAU).
a Histology.
b FIAU quantitative autoradiography (QAR), 24-h study, nonwashed.
c 2-[^{14}C]FIAU QAR, 24-h study, washed

in vivo transduced RG2(TK$^+$) tumor was 0.96% dose/g; adjacent brain activity was below the level of detection on the autoradiograms.

The *second series* of experiments involved s.c. implantation of 10^6 wild-type RG2 cells into both flanks of athymic Harlan Sprague Dawley R-Nu rats (Tjuvajev et al. 1996). Following a 46-day growth period, the right and left flank tumors reached a size of 5×4×3 and 3×2×1 cm, respectively. The larger left-sided tumor was inoculatedwith 10^6 gp-STK-A2 vector-producer cells (retroviral titer, 10^6–10^7 cfu/ml) in 100 μl minimal essential medium (MEM;without fetal calf serum, FCS) to induce in vivo transfection and transduction of the tumor. No carrier-added [^{131}I]FIAU was prepared, and 2.8 mCi was injected i.v. into the

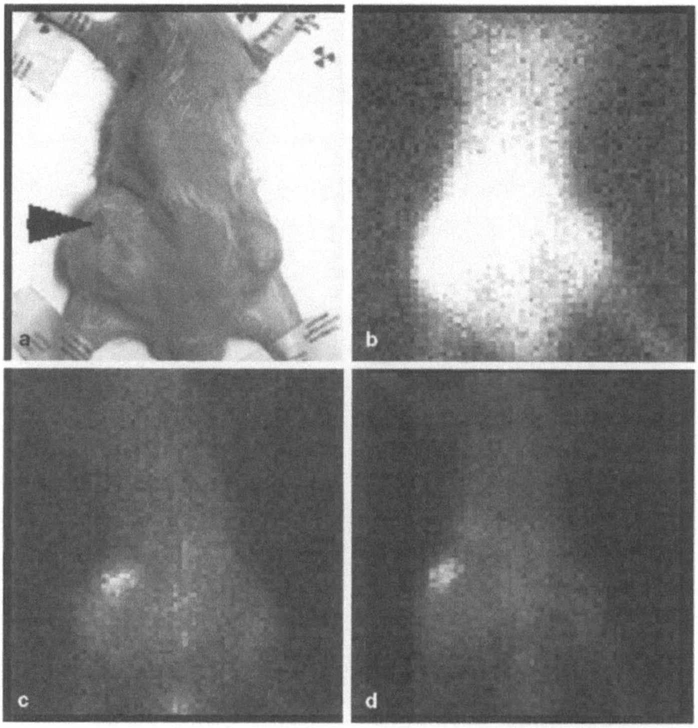

Fig. 8a–d. Gamma camera imaging of HSV1-tk gene transduction and expresssion with 2-[^{14}C]5-iodo-2'fluoro-2'-deoxy-1-β-D-arabinofuanosyl-uracil (FIAU). **a** Transduced (*arrow*) and nontransduced (contralateral) flank tumors. **b** 4-h gamma cameral image. **c** 24-h image. **d** 36-h image normalized to the same level of intensity (radioactivity)

animal (that previously received a thyroid-blocking dose of NaI) 14 days after gp-STK-A2 cell inoculation. Gamma camera imaging was performed at 4, 24, and 36 h after the [^{131}I]FIAU injection with a dual-headed ADAC Genesys gamma camera (Milpitas, CA) equipped with a high-energy, high-resolution collimator. The images were normalized to a reference standard (not shown in the field of view) to account for differences in counting time (10-, 15-, and 20-min period, respectively) and decay; they can be compared to a scaled photographic

Fig. 9a–c. Region of transduction for the tumor (*arrow*) shown in Fig. 8. **a** Quantitiative autoradiography (QAR). **b** Macrohistology. **c** Histology (×40)

image of the animal (the arrow indicates the site of inoculation of vector producer cells into the tumor; Fig 8.). The normalized images demonstrate washout of radioactivity from the body with retention of activity in the area of gp-STK-A2 cell inoculation. The 24- and 36-h images show specific localization of retained radioactivity in the same area where the vector-producer cells were injected. It is apparent that the area of retained radioactivity (area of transduction) was limited and did not extend throughout the tumor. The nontransduced contralateral tumor (right side) and other tissues did not show any retention of radioactivity. These sequential images show the importance of a washout strategy and justify delayed imaging at 24 or 36 h after FIAU administration.

We also measured levels of radioactivity with QAR in the tumor that was visualized by the gamma camera. The tumor was surgically resected after gamma camera imaging (36 h) and prepared for QAR as described above. The QAR images of HSV-TK expression in the transduced tumor were compared to the *same* hematoxylin and eosin-stained section that produced the QAR image (Fig. 9). The QAR, macrohistology and histology (×40) area of HSV-TK expression (high [^{131}I]FIAU activity) in the tumor is clearly seen on the QAR image and reflects the transduced portion of the tumor. This area of tumor transduction can be readily identified because it retains significantly higher levels of radioactivity (~1% dose/g) than other nontransduced areas (~0.03 % dose/g) and reflects the expression of HSV-TK. The area of highly retained radioactivity corresponds to a tumor nodule that was located more peripherally in the tumor and at the site of vector-producer cell injection. Such localized and highly effective transduction may be due to the presence of a thin fibrous capsule which surrounds this tumor node and demarcates it from adjacent tumor tissue. Presumably, the 10^6 gp-STK-A2 vector-producer cells were injected into this tumor nodule, although no producer cells could be identified on morphological criteria. This tumor node, with high levels of the retained radioactivity (white box), had microscopic features typical of subcutaneously growing RG2 glioma. Some of the tumor tissue around this nodule was necrotic. The pattern of HSV1-TK immunohistochemistry corresponded to the pattern of [^{131}I]FIAU radioactivity seen in the autoradiograms (Tjuvajev et al. 1996).

The *third series* of experiments involved s.c. implantation of 10^6 wild-type W256 cells into the dorsum of the neck and right flank and W256TK$^+$ cells into the left flank of athymic Harlan Sprague Dawley

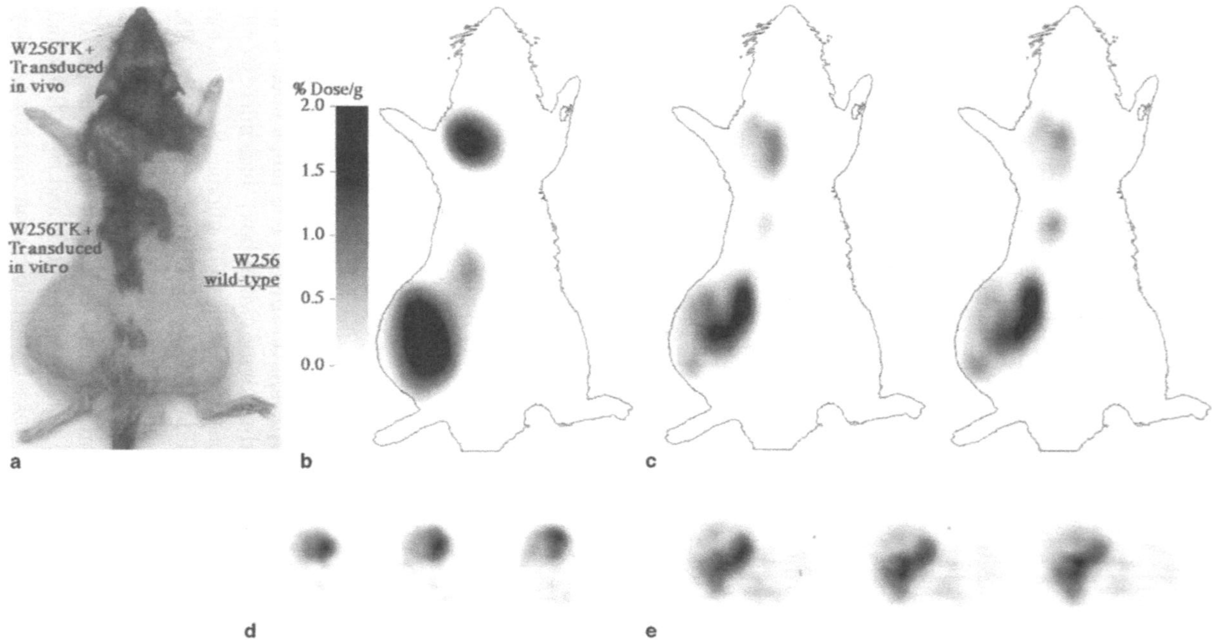

Fig. 10a–e. Positron emission tomography (PET) imaging of HSV1-tk gene transduction and expression using [^{124}I]5-iodo-2'-fluoro-2'-deoxy-1-β-D-arabinofuanosyl-uracil (FIAU). **a** Photograph of tumor-bearing rat. **b** Planar PET image. **c, d** Coronal PET images. **e** Transaxial PET images of through the in vivo transduced neck tumor and the in vitro induced flank tumor

R-Nu rats. Following a 14-day growth period, the flank and cervical tumors reached a diameter of approximately 10 and 5 mm, respectively. The neck tumor (wild type) was inoculated with 10^6 gp-STK-A2 vector-producer cells (retroviral titer, 10^6–10^7 cfu/ml) in 100 μl MEM (without FCS) to induce in vivo transduction of the tumor. No carrier-added [^{124}I]FIAU was prepared, and 25 μCi was injected i.v. into each animal (that previously received a thyroid-blocking dose of NaI), 14 days after gp-STK-A2 cell inoculation. PET imaging was performed at 30 and 50 h after the [^{124}I]FIAU injection with a General Electric Advance PET tomograph (Milwaukee, WI). The PET acquisitions were obtained over 30 and 60 min, respectively, in three-dimensional (septa-out) mode. A transmission scan was performed prior to each emission acquisition, and the reconstructed images were corrected for attenuation and scatter. A planar image (summated coronal images) along with two coronal image planes through the flank tumors is shown in Fig. 10; the images were acquired 50 h after [^{124}I]FIAU administration. Three transaxial image planes are also shown through the neck and flank tumors, respectively.

Localization of radioactivity to the in vitro transduced (left flank) and in vivo transduced (neck) tumors is clearly seen, but only low background levels of radioactivity were observed in the nontransduced wild-type W256 tumor in the right flank. Also note the heterogenous pattern of radioactivity distribution in both transduced tumors, demonstrating the nonclonal origin of the W256TK$^+$ cells (bulk culture) used to produce the left flank tumor and the nonuniform transduction of the neck tumor following intratumoral injection of gp-STK-A2 vector-producer cells. The grey scale bar pertains to the transaxial images only.

11.10 Feasibility of Clinical Applications Using PET and SPECT Imaging

The feasibility of imaging HSV1-TK expression in patients is dependent upon several factors:

1. The level of HSV1-TK expression in the transduced tissue to be imaged
2. The sensitivity and selectivity of the marker substrate
3. The administered dose and clearance of the marker from the blood

4. The sensitivity of the imaging equipment
5. Scan duration

We previously addressed the feasibility of PET and SPECT imaging of "clinically relevant" levels of HSV-TK expression is some detail, based on our preliminary in vitro and in vivo data, the anticipated dose of radiolabeled 2'-fluoro nucleoside (Bq) administeredfor PET and SPECT studies, the clearance of 2'-fluoro nucleoside from the blood, and the sensitivity of our SPECT and PET instruments. We have calculated that the level of [^{124}I]FIAU radioactivity at 24 h in transduced tumor tissue expressing "clinically relevant" levels of HSV-TK will be approximately 0.07–0.12 μCi/cc for a 2-mCi dose (see Appendix). This is approximately 100 and 500 times higher for septa-in and septa-out PET acquisitions, respectively, than the lower limit sensitivity of the PET tomograph. For [^{131}I]FIAU and [^{123}I]FIAU SPECT acquisitions following a 10- and 30-mCi dose of FIAU, respectively, we estimate a count rate that is only approximately five- and tenfold above the lower limit sensitivity of a dual-headed SPECT tomograph.

11.10.1 Safety Issues

The issue of 2'-fluoro nucleoside toxicity and the potential for toxicity from administering radiolabeled 2'-fluoro nucleosides for diagnostic studies has been raised in some quarters. This concern relates to the serious complications and two deaths that were associated with a clinical trial evaluating the long-term administration of FIAU as a potential antiviral agent in treatment of patients with chronic hepatitis B virus infection (McKenzie et al. 1995) and to the ethics of administering tracer doses of FIAU and other 2'-fluoro nucleoside radiopharmaceuticals to patients for diagnostic purposes. To appreciate the significance of these issues, the difference between a therapeutic dose of FIAU administered in clinical trials and an equivalent dose administered for diagnostic imaging studies must be clearly understood. A comparison between the "dose administered" in clinical trials to the "equivalent dose" of radiolabeled 2'-fluoro nucleosides that would be administered in a diagnostic PET study was presented previously (Tjuvajev et al. 1996). It should be noted that a no-carrier-added procedure for radioiodination of

FIAU (and other 2'-fluoro nucleosides) has been developed (Misra et al. 1986) and was the radiosynthetic procedure used for labeling FIAU. It is also the method proposed for labeling other 2'-fluoro nucleosides (such as FIVAU and FIVRU) with [^{131}I], [^{123}I], and [^{124}I] (Misra et al. 1986). The mass of radiopharmaceutical produced from a no-carrier-added synthesis can be calculated using Avogadro's number, the half-life of the radionuclide (in seconds), the molecular weight of the radiopharmaceutical, and the administered dose (in Bq). For example, the equivalent mass of a 74-MBq (2-mCi) dose of no-carrier-added [^{124}I]FIAU is only 16 ng (proposed for future PET imaging studies in patients); for a 1.1-GBq (30-mCi) dose of no-carrier-added [^{123}I]FIAU, the equivalent mass of FIAU that would be administered is only 32 ng (proposed for future SPECT imaging studies). A dose of 16 ng FIAU represents only 1/220000–1/7440000 of the *daily* dose (and 1.5×10^{-5} to 8.2×10^{-8} of the *total* dose) that was administered to patients in the 14- to 28-day hepatitis B clinical trials, in which *no* FIAU-related toxicity was observed (McKenzie et al. 1995). Only in the 168-day clinical trial (H3X-PPPC) did severe liver and pancreatic toxicity and neurotoxicity appear.

11.10.2 Dosimetry

The radiation exposure from diagnostic studies of HSV-tk expression are expected to be small and well within acceptable limits. This is certainly true for the [^{18}F]-labeled compounds where maximum doses of 370 MBq (10 mCi) would be expected to be administered in patient studies. In the case of [^{123}I]-, [^{124}I]-, or [^{131}I]-labeled compounds, the physical half-life of the isotopes are longer (13 h, 4 days, and 8 days, respectively). The dosimetry of iodine-labeled 2'-fluoro nucleosides to critical organs would be expected to be considerably less than that of IUdR, which is currently undergoing clinical evaluation as an imaging agent for tumor proliferation (Tjuvajev et al. 1994; Blasberg et al. 1996). This is because the bulk of the radioactivity would be excreted in the urine as unchanged parent compound, due to the stability and resistance to metabolic degradation of this class of compounds (Table 1). The major radiolabeled metabolite is expected to be iodide, although the fraction and amount of radiolabeled iodide in blood and tissue is expected to be much less compared to that following [*I]IUdR administration. Furthermore, the 2'-fluoro nucleosides are not (or only minimally) incorporated into DNA of non-HSV1-tk-transduced replicating cells, as is IUdR (see Figs. 1–3). Thus radiation (Auger electron) exposure due to 2'-fluoro nucleoside incorporation into DNA of nontransduced proliferating tissues (e.g., bone marrow and intestinal mucosa) is expected to be small and substantially less than that of IUdR. Similarly, the accumulation of radiolabeled iodide in thyroid and gastric mucosa (expressed as a percentage dose per gram) is expected to be substantially less than that following administration of [*I]IUdR.

11.11 Conclusions

We have demonstrated a "proof of principle," namely, that the expression of successful gene transduction can be monitored in vivo using noninvasive imaging techniques. This was demonstrated by QAR, gamma camera, SPECT, and PET imaging techniques. It was accomplished by selecting an appropriate marker gene and marker substrate combination based on in vitro experiments performed in tissue culture and involved several different experimental animal models. HSV1-tk was chosen as a model marker gene, and radiolabeled FIAU was shown to be a substantially better marker substrate for the HSV1-TK enzyme than IUdR or ganciclovir. The magnitude of FIAU accumulation in different RG2TK^{+} transduced clones was highly correlated with two independent measures of HSV1-tk expression, namely, their sensitivity to ganciclovir and the level of HSV-tk mRNA expression. The 2'-fluoro nucleoside analogues of uracil have specific advantages as marker substrates for HSV1-tk expression. This is primarily due to the following: (a) their selective phosphorylation by HSV1-TK (as compared to endogenous mammalian thymidine kinase), (b) their resistance to metabolic degradation by nucleoside phosphorylase, (c) their retention in the tissue during the period of imaging (due to their phosphorylation and incorporation into DNA), and (d) the absence (or low levels) of radiolabeled metabolites at the time of imaging (due to their rapid clearance from the body and the long half-life of the iodine radionuclides used for labeling). These four advantages eliminate (or markedly reduce) the problems associated with imaging radiolabeled metabolites in both target and surrounding tissue, and they facilitate quantitative measurements and interpretation of the resultant images.

Appendix

Feasibility of Imaging Clinically Relevant Levels of HSV1-tk Expression

If we can identify the "minimum" level of HSV-TK expression in targeted tissue (e.g., tumor) necessary for therapeutic efficacy, then we can define the range of HSV-TK expression that would be clinically relevant for imaging. In an attempt to estimate the minimum level of HSV-TK expression necessaryfor clinical efficacy and to estimate whether that level of HSV-TK expression could be imaged with FIAU, we have used data presented in Figs. 1–5 and made a number of assumptions. First, we will assume that therapeutic efficacy is related to both the level of ganciclovir maintained in the blood and the level of HSV-tk expression in the targeted tumor tissue. The blood levels of ganciclovir that are expected to be achieved in clinical trials following i.v. administration of 5 mg/kg q 12 h (when creatinine clearance is 50 ml/min per /1.73 m^2; Sandoz HSV-TK gene therapy clinical trials) are 1–10 μg/ml plasma. We will also assume that this blood level is maintained relatively constant and that transduced human tumors and transduced RG2 cells have similar sensitivities (ED50) to ganciclovir. For example, if a level of ganciclovir of 1 μg/ml plasma corresponds to a concentration of 1 μg/ml in the culture medium and if 1 μg/ml is the ED_{50} concentration of ganciclovir (e.g., the concentration necessary to achieve a therapeutic effect), then the minimum level of HSV-TK expression in targeted tumor tissue necessary to achieve a therapeutic effect could be estimated from the plot shown in Fig. 5b. This ED_{50} concentration of ganciclovir (1 μg/ml) corresponds to the level of HSV-TK expression in $RG2TK^+$ cells. This level of HSV-TK expression in tumor cells may represent a "minimum level" of expression that would be necessary to achieve a therapeutic response in clinical studies (corresponding to the level of HSV-TK expression in $RG2TK^+$ cells that is necessary for an ED_{50} response when the ganciclovir concentration is 1 μg/ml; Fig. 5b). Thus we have an approximate value as the answer of the first question: *What is the minimum level of HSV-TK (enzyme) expression necessary for therapeutic efficacy?*

The following comparisons and calculations address the second question: *Can "clinically relevant" levels of HSV-TK expression be* *selectively imaged with [^{124}I]FIAU and PET (and possibly with [^{131}I]- or [^{123}I]FIAU and SPECT)?* The clinically relevant, "minimum" level of HSV-TK expression estimated above correspond to the level of HSV-TK expression in RG2TK$^+$ cells (Fig. 5b). For in vivo experiments in rats, we can calculate the FIAU accumulation rate (plasma clearance, K_i) from the data shown in Figs. 6–10 and a relationship previously used by the author (Molnar et al. 1984). The accumulation rate (plasma clearance) constant of FIAU is approximately 10 µl/min per g in RG2TK$^+$ brain tumors, approximately 1.2 µl/min per g in nontransduced RG2 tumors, and .3 µl/min per g in remote brain tissue (see Fig. 6). To estimate the level of radioactivity in transduced tumor tissue of patients expressing HSV-TK at a "minimum" level (i.e., that necessary to obtain a therapeutic effect at a plasma ganciclovir concentration of 1 µg/ml), we used the clearance constant, K_i, of FIAU accumulation in RG2TK$^+$ brain tumors in rats (approximately 0.010 ml/min per g) multiplied by the FIAU input function integral (11763 min-nCi/cc) calculated from FIAU blood clearance measurements in patients (Feinberg et al. 1985) for an injected dose of 2 mCi [^{124}I]FIAU. This calculation yields a value of approximately 0.12 µCi/cc [^{124}I]FIAU at 24 h in tumor tissue expressing "minimally" effective levels of HSV-TK. The background activity in nontransduced tumor is estimated to be approximately 0.014 µCi/cc, and that in normal brain as .003 µCi/cc. These calculations are summarized in the Table A1.

This table includes estimated values of FIAU radioactivity in transduced patient tumors expressing two different levels of HSV-TK that are likely to be clinically significant, namely, the accumulation of FIAU radioactivity at the "minimal" therapeutically effective level of HSV-TK expression when plasma ganciclovir levels are 1.0 and 10 µg/ml plasma, respectively.

In support of the potential for PET (and possibly SPECT) imaging of HSV-TK expression, we have determined the sensitivity of the instruments available to us and calculated the count rate which we expect to see by using [^{124}I]-labeled FIAU and PET imaging, as well as that expected with [^{131}I]- and [^{123}I]FIAU SPECT imaging. Conventional two-dimensional PET imaging of [^{124}I] has a count rate 22 times that achieved with [^{131}I] and SPECT (Table A2); by using three-dimensional (septaless) PET acquisitions, there will be an improvement in count rate by a factor of 116. It should be noted that substantial improvements are

Table A1. Feasability of imaging HSV1-tk gene expression in patients: calculations based on FIAU data in tissue culture and in animals

GCV concentration (μg/ml)[a]	HSV-TK mRNA[b,c]	FIAU K_i[d] (μl/min per g)	Estimated FIAU activity at 24 h in transduced patient tumors (μCi/cc)
1.0	32	~10[e]	~0.12[g]
10	19	~6[f]	~0.07[h]

GCV, ganciclovir; FIAU, 5-iodo-2'-fluoro-2'-deoxy-1-β-D-arabinofuranosyl-uracil.

[a]Tissue culture ED_{50} levels for calculating HSV-TK mRNA; patient plasma levels in the Sandoz gene therapy clinical trial.

[b]HSV-TK mRNA levels in transduced cells at ED_{50} concentrations of ganciclovir in the culture medium (see Fig. 5).

[c]HSV1-TK mRNA/S28 RNA IOD.

[d]Plasma clearance constant.

[e]Calculated from in vivo experiments.

[f]Calculated from in vivo and in vitro data corrected for the lower expression of HSV-TK mRNA.

[g]For patient tumors expressing HSV-TK at a mRNA/S28 RNA ratio of 32, a K_i of 9.9 μl/min per g, and a 2-mCi dose of FIAU.

[h]For patient tumors expressing HSV-TK at a mRNA/S28 RNA ratio of 19, a K_i of 6.0 μl/min/g, and a 2-mCi dose of FIAU.

achieved despite a fivefold difference in injected dose (2 vs, 10 mCi), a 77% reduction in detectable emissions (only 23% of ^{124}I decay is β^+), and a shorter half-life for [^{124}I] compared to [^{131}I] (4 vs. 8 days). Table A2 also has a column indicating the lower limit of measurable radioactivity (μCi/cc) within a 1-cc homogeneous tumor, which can be accurately measured (10% error) in a 1-h scan with each of the modalities.

Although these calculations are approximations, they suggest that the level of HSV-tk expression that is necessary for therapeutic efficacy is more than sufficient for imaging with [^{124}I]FIAU and PET. The possibility of imaging comparable levels of HSV-TK expression with [^{131}I]FIAU or [^{123}I]FIAU and SPECT is less certain.

An alternative indication that good-quality PET images of HSV-tk expression in transduced tissue should be obtained at 24 h following i.v.

Table A2. Tomograph sensitivity and dose of radiolabeled FIAU projected for patient studies

Tomograph/ radionuclide	Sensitivity (cpm/μCi)	Dose (mCi)	Relative no. of detected events at 24 h	Lower limit of measurable activity (μCi/cc)
SPECT [^{131}I] HEHR[a]	27.9	10	1.0	0.1
SPECT [^{123}I] LEHR[a]	84.4	30	1.9	0.036
PET [^{124}I] GE ADvance HS[b]	2872	2	22	0.0011
PET [^{124}I] GE Advance 3D[b]	15092	2	116	0.0002

FIAU, 5-iodo-2'-fluoro-2'-deoxy-1-β-D-arabinofuranosyl-uracil; SPECT, single photon emission computed tomography; PET, positron emission tomography.

[a]Dual-headed ADAC Genesys gamma camera (Milpitas, CA) equipped with high-energy, high-resolution (HEHR) and low-energy, high-resolution (LEHR) collimators for [^{131}I] and [^{123}I], respectively.

[b]General Electric Medical Systems, Milwaukee, Wisconsin, USA; standard "septa-in" high-sensitivity [HS] mode; "septa-out" three-dimensional (3D) mode.

administration of 2 mCi (74 MBq) is given by a comparison of fluorodeoxyglucose (FDG) and FIAU activity in rat brain and transduced tumors, respectively. Radioactivity (expressed as percentage dose per gram tissue) in the frontal cortex of conscious rats, 60 min after i.v. injection of FDG, was 1.68±0.38 (n=25) (unpublished data from our laboratory). For tumors expressing HSV-tk in rats, radioactivity levels 24 h after i.v. injection of FIAU ranged between 0.1% and 1.0% dose/g tissue. Thus imaging HSV-tk gene expression in transduced tissue with [^{124}I]FIAU and PET is feasible, particularly if image acquisition lasts 30 min or more and is performed using high-sensitivity, "septa out" technology available on the newer PET tomographs.

References

Alauddin MM, Conti PS, Lever JR, Mazza SM, Hamzeh FM (1996) Synthesisof F-18 9-[(3-fluoro-1-hydroxy-2-propoxy)-methyl]-guanine (FHPG) for in vivo imaging of viral infection and gene therapy with PET. J Nucl Med 37:5,193P

Blasberg RG, Roelcke U, Weinreich R, von Ammon K, Crompton N, Guenther I, Knust E, Beattie B, Finn R, Leenders K (1996) [124I]-Iododeoxyuridine imaging tumor proliferation. J Nucl Med 37:5, 54P

Blasberg RG, Groothuis D, Molnar P (1981) Application ofquantitative autoradiographic measurements in experimental brain tumormodels. Semin Neurol 1:203–223

Blasberg RG, Gazendam J, Patlak CS, Fenstermacher JD (1980) Quantitative autoradiographic studies of brain edema and acomparison of multi-isotope autoradiographic techniques. In: Cervos-Navarosand J, Ferszt S (eds) Brain edema. Raven, New York, pp 255–270 (Advances in neurology, vol 28)

Blum MR, Liao SH, deMiranda P (1982) Overview of acyclovir pharmacokinetic disposition in adults and children. Am J Med 73:197–201

Borrelli E, Heyman R, Hsi M, Evans RM (1988) Targetingof an inducible toxic phenotype in animal cells. Proc Natl Acad SciUSA 85:7572–7576

Clinical Protocol List (1995) Cancer Gene Ther 2 (3):225

Culver KW, Blaese RM (1994) Gene therapy for cancer. Trends in Genet 10:174–178

Culver KW, Ram Z, Wallbridge S, Ishii H, Oldfield EH, Blaese RM (1992) In vivo gene transfer with retroviral vector-producercells for treatment of experimental brain tumors. Science 256:1550–1552

Feinberg A, Leyland-Jones B, Fanucchi MP, Hancock C, Fox JJ, Watanabe KA, Vidal PM, Williams L, Young CW, Philips FS (1985) Biotransformation and elimination of 2-[^{14}C]-5-iodo-(2'-fluoro-2'-deoxy-1-β-D-arabinofuranosyl)-cytosine in immunosuppressed patients with herpesvirus infections. Antimicrob Agents Chemother 27(5):733–738

Fletcher C, Sawchuk R, Chinnock B, deMiranda P, Balfour HH Jr (1986) Human pharmacokinetics of the antiviral drug DHPG. Clin Pharmacol Ther 40:281–286

Ghattas IR, Sanes JR, Majors JE (1991) The encephalomyocarditisvirus internal ribosome entry site allows effieicnt coexpression of two genes from a recombinant provirus in cultured cells and in embryos. Mol Cell Biol 11:5848

Haberkorn U, Altmann A, Morr I, Germann C, van Kaick G. (1996) Uptake of specific substrates for HSV thymidine kinase in genetically modifiedhepatoma cells. J Nucl Med 37:5, 235P

Haberkorn U, Oberdorfer F, Gebert J, Morr I, Haack K, Weber K, Lindauer M, van Kaick G, Schackert HK (1996) Monitoring gene therapy with cytosine deaminase: in vitro studies using tritated-5-fluorocytosine. J Nucl Med 37: 87–94

Iwashina T, Tovell DR, Xu L, Tyrrell DL, Knaus EE, Wiebe LI (1988) Synthesis and antiviral activity of IVFRU, a potential probe for the non-invasive diagnosis of herpes simplex encephalitis. Drug Design Deliv 3:309–321

Jolly D (1994) Viral vector systems for gene delivery. Cancer Gene Therapy 1(1):51–64

Ko L, Koestner A, Wechsler W (1980) Morphological characterisation of nitrosurea-induced glioma cell lines and clones. Acta Neuropathol (Berl) 84:529–541

Kong XB, Vidal P, Tong WP, Chiang J, Gloff CA, Chou TC (1992) Preclinical pharmacology of the anti-hepatitis virus agent 2'fluoro-5-ethyl-1beta-D-arabinofuranosyluracil in mice and rats. Antimicrob Agents Chemother 36:1472–1477

Manthorpe M, Cornefert-Jensen F, Hartikka J, Felgner J, Rundell A, Margalith M, Dwarki V (1993) Gene therapy by intramuscular injection of plasmid DNA: studies on firefly luciferase gene expression in mice. Human Gene Ther 4(4):419–431

McKenzie R, Fried MW, Sallie R et al (1995) Hepatic failureand lactic acidosis due to fialuridine (FIAU), an investigational nucleoside analogue for chronic hepatitis B. N Engl J Med 333:1099–1105

Misra HK, Knaus EE, Wiebe LI, Tyrell DL (1986) Synthesis of ^{131}I, ^{125}I, ^{123}I and ^{82}Br labelled 5-halo-1-(2-deoxy-2-fluoro-β-D-arabinofuranosyl) uracils. Appl Radiat Isot 37:901–905

Molnar P, Groothuis D, Blasberg R, Zaharko D, Owens E, Fenstermacher J (1984) Regional thymidine transport and incorporation in experimental bràin and subcutaneous tumorsJ Neurochem 43(2):421–432

Moolten FL (1986) Tumor chemosensitivity conferred by inserted herpes thymidine kinase gene: paradigm for a prospective cancer control strategy. Cancer Res 46: 5276–5281

Moolten FL, Wells JM (1990) Curability of tumors bearing herpes thymidine kinase genes transfected by retroviral vectors. J Natl Cancer Inst 82(4):297–300

Mullen CA, Kilstrup M, Blase M (1992) Transfer of the bacterial gene for cytosine deaminase to mammalian cells confers lethal sensitivity to 5-fluorocytosine: a negative selection system. Proc Natl Acad Sci USA 89:33–37

Mulligen R, Berg P (1981) Selection for animal cells that express the Escherichia coli gene coding for xanthine-guanine phosphoribosyl transferase. Proc Natl Acad Sci USA 78:2072–2076

Price J, Turner D, Cepko C (1987) Lineage analysis in the vertebrate nervous system by retrovirus-mediated gene transfer. Proc Natl Acad Sci USA 84:156–160

Price R, Cardle K, Watanabe K (1983) The use of antiviral drugs to image herpes encephalitis. Cancer Res 43: 3619–3627

Saito Y, Price RW, Rottenberg DA, Fox JJ, Su T-L, Watanabe KA, Philips FA (1982) Quantitative autoradiographic mappingof herpes simplex virus encephalitis with radiolabeled antiviral drug. Science 217:1151–1153

Saito Y, Rubenstein R, Price RW, Fox JJ, Watanabe KA (1984) Diagnostic imaging of herpes simplex virus encephalitis using radiolabeled antiviral drug: autoradiographic assessment in animal model. Ann Neurol 15:548–558

Sokolic RA, Sekhsaria S, Sugimoto Y, Whiting-Theobald N, LintonGF, Li F, Gottesman MM, Malech HL (1996) A bicistronic retrovirus vector containing a picornavirus internal ribosome entry site allows for correction of X-Linked CGD by selection for *MDR*1 expression. Blood 87(1):42–50

Srinivasan A, Gambhir SS, Green AL et al (1996) A PET reporter gene (PRG)/PET reporter probe (PRP) technology for repeatedly imaging gene expression in living animals. J Nucl Med 37:5, 107P

Sugimoto Y, Aksentijevich I, Murray GJ, Brady RO, Pastan I, Gottesman MM (1995a) Retroviral coexpression of a multidrug resistance gene (*MDR*1) and human alpha-galactosidase A for gene therapy of Fabry disease. Human Gene Ther 6:905–915

Sugimoto Y, Hrycyna CA, Aksentijevich I, Pastan I, Gottesman MM (1995b) Coexpression of a multidrug-resistance gene (MDR1) and herpes simplex virus thymidine kinase gene as part of a bicistronic messenger RNA in a retrovirus vector allows selective killing of MDR1-transduced cells.Clin Can Res 1:447–457

Tjuvajev JG, Muraki A, Ginos J, Berk J, Koutcher J, Ballon D, Beattie B, Finn R, Blasberg R (1993) Iododeoxyuridine uptake and retention as a measure of tumor growth. J Nucl Med 34:1152–1162

Tjuvajev JG, Macapinlac HA, Daghighian F, Scott AM,Ginos JZ, Finn RD, Kothari P, Desai R, Zhang Z, Beattie B,Graham M, Larson S, Blasberg RG (1994) Imaging of brain tumor proliferativeactivity with iodine-131-iododeoxyuridine. J Nucl Med 35:1407–1417

Tjuvajev J, Stockhammer G, Desai R, Uehara H, Watanabe K, Gansbacher B, Blasberg R (1995) Imaging the expression of transfected genes in vivo. Cancer Res 55:6126–6132

Tjuvajev JG, Finn R, Watanabe K, Joshi R, Oku T, Kennedy J, BeattieB, Koutcher J, Larson S, Blasberg RG (1996) Noninvasive imaging of herpesvirus thymidine kinase gene transfer and expression: a potential methodfor monitoring clinical gene therapy. Cancer Res 56:4087–4095

Tovell D, Yacyshyn H, Misra H, Knaus EE, Wiebe LI, Samuel J, Gill MJ, Tyrrell LJ (1987) Effect of acyclovir on the uptake of ^{131}I-labeled 5-iodo-(2'-fluoro-2'-deoxy-1-β-D-arabino-furanosyl)-uracil in herpes infected cells. J Med Virol 22:183–188

12 Intra- and Intercellular Recognition of Cell Signaling Using Oligonucleotides

M.K. Dewanjee

12.1 Introduction 202
12.2 Roles of Cytokines in Treatment of Inflammation, Infection, and Septic Shock 223
12.3 Noninvasive Diagnostic Imaging with Radiopharmaceuticals (Tracers) 224
12.3.1 Nucleic Acid-Based Radiolabeled Probes 226
12.3.2 Hybridization of Radiolabeled Oligonucleotide Probe with mRNA 227
12.3.3 Rational Development of Nucleic Acid-Based Radiopharmaceuticals 228
12.3.4 Mechanism of Inhibition by Antisense Oligonucleotide 229
12.4 Synthesis, Conjugation, Radiolabeling, Purification, and Characterization of Aminohexyl-Conjugated Oligodeoxynucleotides 230
12.5 Design of Novel and Conventional Radiopharmaceuticals 233
12.6 Structure and Function of Gene-Regulatory Proteins 234
12.7 Protein–Small Molecule, Protein–Protein, and Protein–DNA Interactions in the Signaling Pathways for Transcription, Translation, and Cell Cycle Progression 236
12.8 Apoptosis Program for Treatment of Cancer and Restenosis After Angioplasty 237
12.9 Cell Signaling for Transcription, Translation, and DNA Replication 238
12.10 Protein Kinase Superfamily 241
12.10.1 Inhibitors of Protein Kinase A 242
12.10.2 Linkage of Receptor Protein Kinases to Ras Protein Superfamily and Downstream Mitogen-Associated Protein Kinases 243

12.10.3 Activation of Mitogen-Associated Protein Kinase Cascades by Ras ... 243
12.11 Cell Signaling in Cell Cycle and DNA Replication ... 244
12.12 Restenosis, Endothelial Damage and Early Gene Response ... 245
12.13 Dependence of Tyrosine Kinase-Associated Receptors on Nonreceptor Tyrosine Kinases ... 245
12.14 Interaction Between Antisense Oligonucleotides and mRNA Targets ... 249
12.14.1 Molecular Mechanism ... 249
12.14.2 Parameters of Affinity in Oligonucleotide–mRNA Hybridization Reactions ... 250
12.14.3 Specificity of Hybridization Reactions ... 250
12.15 Potential Application of Ribozyme in Inhibition of Transcription 251
12.16 Applications of Phosphorothioate Oligonucleotides in Diagnosis and Therapy ... 251
12.17 Chemistry of Modified Oligonucleotides ... 252
12.18 Adverse Effects of Administration of Oligonucleotides ... 252
References ... 254

12.1 Introduction

Cell growth, normal development, and diseases depend on interactions among groups of small molecules and macromolecules of structural and functional proteins and nucleic acids. Bucket brigades of several essential groups of signaling proteins carry messages from outside the cell membrane to the nucleus in order to make new mRNAs and proteins, new DNA, or new cells, to repair DNA, or to die by falling apart (apoptosis). This controlled process of cell growth and development is circumvented by several gene mutations in cancer cells. Mutated genes generate critical signaling proteins and make cancer cells independent of other cells and other normal proteins for rapid growth in benign or malignant tumors.

This review attempts to identify mRNA targets in the multiple signaling pathways and cites the recent literature concerning ingenious applications of oligonucleotides and their stable analogues in several categories (Agrawal 1991; Akhtar and Agrawal 1997; Blatteis 1997; Egholm et al. 1993; Fenster et al. 1994; Holt 1996; Kedes and Kedes 1996; Monia et al. 1992; Gewirtz et al. 1996; Stein and Cheng 1993; sun and Helene

1993). Due to lack of space, individual studies have not been discussed in the text. Information on the major themes of cell signaling, signal processing, structures of oligonuclide derivatives, and methods of radiolabeling is provided in Figs. 1–14. The diversity of targets and the multiple interactions, applications, adverse effects, advantages, and disadvantages of oligonucleotide–peptide- and protein-based radiopharmaceuticals are shown in Tables 1–5.

Molecular biology provides several powerful tools for understanding the etiology of cancer and for diagnosis and therapy. In cancer diagnosis, topics include (a) control of expression of cancer genes, (b) cloning of genes and determining the sequence of nucleotides, (c) identifying oncogenes, (d) determining how they are derived from proto-oncogenes by mutations with virus, chemicals, and radiation, (d) role of proteins made by oncogenes, i.e., oncoproteins, (e) survival of cancer cells with a lower level of growth factors (autonomy of oncoproteins), (f) measurement of mRNA by making radiolabeled "molecular velcro" for amplified oncogenes, and (g) application of the nuclear imaging technique to diagnose amplified mRNA in cancer cells of live animals. We hope that these novel, noninvasive diagnostic procedures will help to achieve early diagnosis in cancer patients in the near future. This methodology could also be extended for therapy. Figures 1–6 explain the principles behind the DNA-based diagnostic strategy.

Specific interaction between detective proteins (antibody) and target protein (antigen) and between specific detective nucleic acid and target nucleic acid (double-stranded DNA or single-stranded mRNA) forms the basis of in vitro and in vivo diagnostic assays (Fig. 12, Tables 1, 2). Depending on the nature of the immunohistochemical method and the type of label used, the reaction products of antibody–antigen complex or of nucleic acid probe–DNA/mRNA complex are visualized by radiation (radioimmunoassay in in vitro systems and radioimmunoscintigraphy in in vivo systems using penetrating radiation), light (chemiluminescence), and fluorescent or electron microscopy.

Pioneering studies performed by Rous (1911) and Avery et al. (1944) indicated that DNA is the principal chemical in the gene that is responsible for the transformation of normal cells by transfection. Crystallographic studies carried out by Watson and Crick (1953) showed that the two complementary strands of DNA are held together by multiple H-bonds between bases. Khorana and colleagues (Narang et al. 1967)

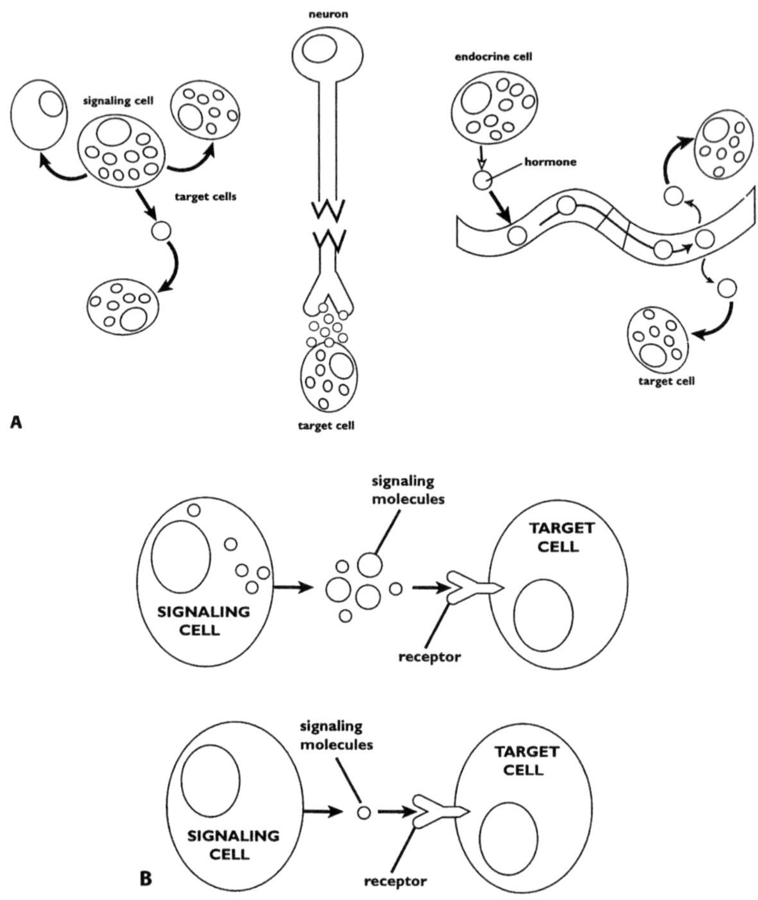

Fig. 1A–C. Different types of cell signaling. **A** Three types of intercellular signaling of paracrine (*left*), synaptic (*center*), and endocrine (*right*) mechanisms. **B** Signaling by secreted molecule (*top*) and by plasma membrane-bound molecule (*bottom*). **C** Signaling at cell membrane, cytoplasm, and nucleus by cell surface receptors (*top*) and intracellular receptors (*bottom*)

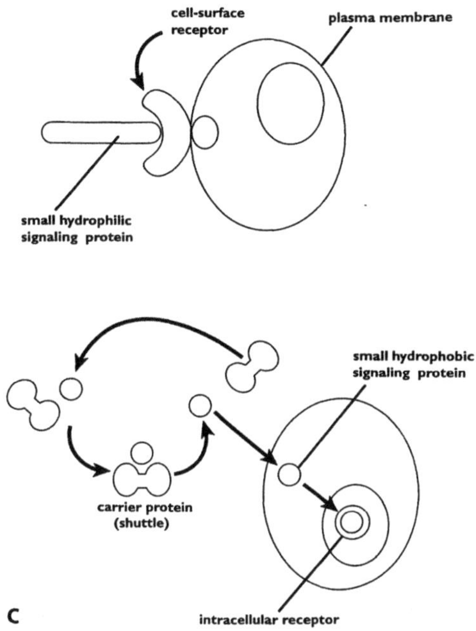

Fig. 1C. Legend see p. 204

demonstrated that DNA could be synthesized by conventional synthetic methods to form phosphodiester linkage with the appropriate four bases of adenine (A), thymine (T), guanine (G), and cytosine (C). Currently, synthetic procedures for making oligonucleotides (100–150 bases) are automated using the solid phase of controlled pore glass beads. Maxam and Gilbert (1977) and Sanger (1981) have developed techniques to determine the sequence of nucleotides of genes by chemical cleavage or aborted synthesis of DNA.

In vivo carcinogenesis is a multistep process. Most of the oncogenes involved are either acquired or derived from mutations of proto-oncogenes induced by chemicals, radiation, and viruses (Stehelin et al. 1976). These oncogenes work synergistically to synthesize oncoproteins (Fig. 9), which continuously generate signals and stimulate the affected cells for proliferation.

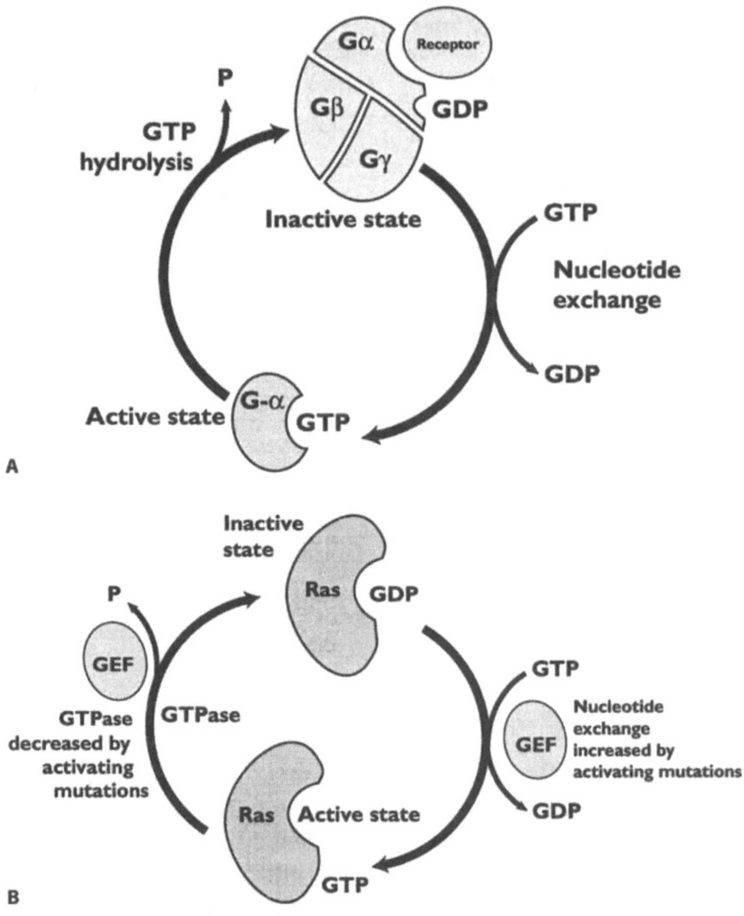

Fig. 2A,B. Signaling by G proteins and guanine nucleotides in cytoplasm of **A** normal cells and **B** mutated cells. Association of GTP rather than GDP with the GTP-binding protein is similar to having an extra phosphate group by phosphorylation

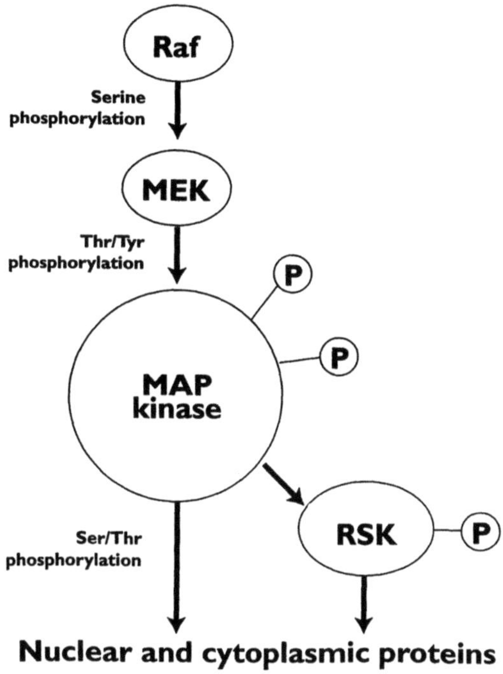

Fig. 3. Signaling pathway for transcription and proliferation via the mitogen-activated protein (MAP) kinase pathway. The number of phosphate additions per transcription and/or DNA replication in a normal cell may be lower than that of a cancer cell (eight to ten vs. ten to 14), since the latter stays in an active, energetically advantageous state and manages to bypass some of the signaling pathways, thus avoiding the multiple checkpoints developed in a normal cell

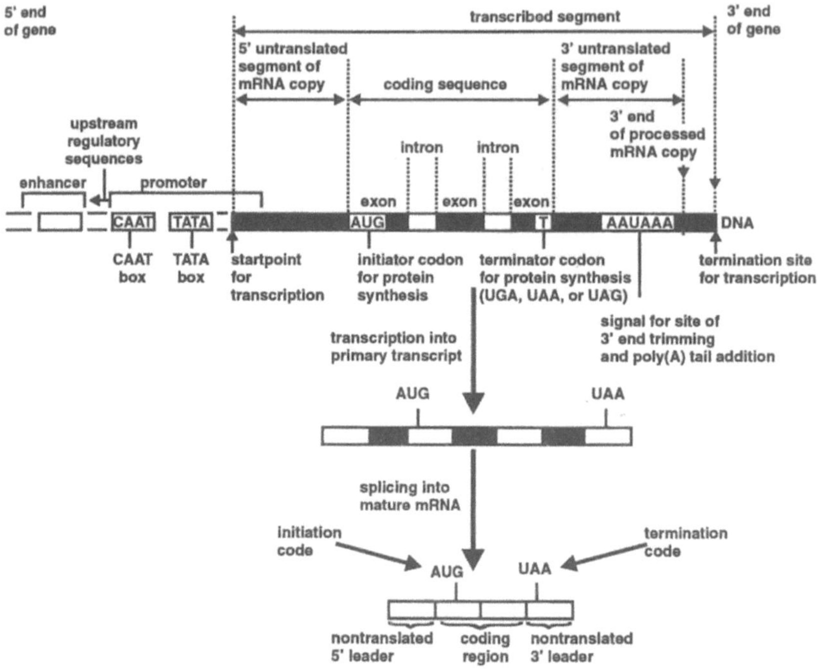

Fig. 4. Synthesis of mRNA and protein from DNA template in a cell. Arrangement of sequence elements or nucleotide blocks in genes coding eukaryotic mRNA molecules. Exons and introns are shown as *boxes*. A gene (less than a thousand to over a million base pairs in size) is preceded by a promoter region and is followed by a terminator region. The start of transcription defines the beginning of the first exon (message of DNA segment exited from nucleus). A gene may contain many exons and introns. Splicing of intron is very precise down to the single nucleotide level to ensure that the joined exons contain the correct reading frame throughout the entire coding region. The final mature mRNA often contains nontranslated regions, called leaders, on its 5' and 3' ends. Translation is always initiated at an AUG codon and terminated by UGA, UAA, or UAG codon. By convention, the sequence elements in DNA are given in their RNA equivalents (U for T), except for the CAAT and TATA boxes, which are given in the same form in which they appear in DNA antisense sequence. The promoter elements contain the CAAT and/or TATA box. There may be one or more introns in the coding region of gene. The enhancer block lying further upstream of the promoter element may contain several repeated blocks of pre-rRNA genes with intergenic spacers. The nucleotide number is positive from downstream of the transcription initiation site

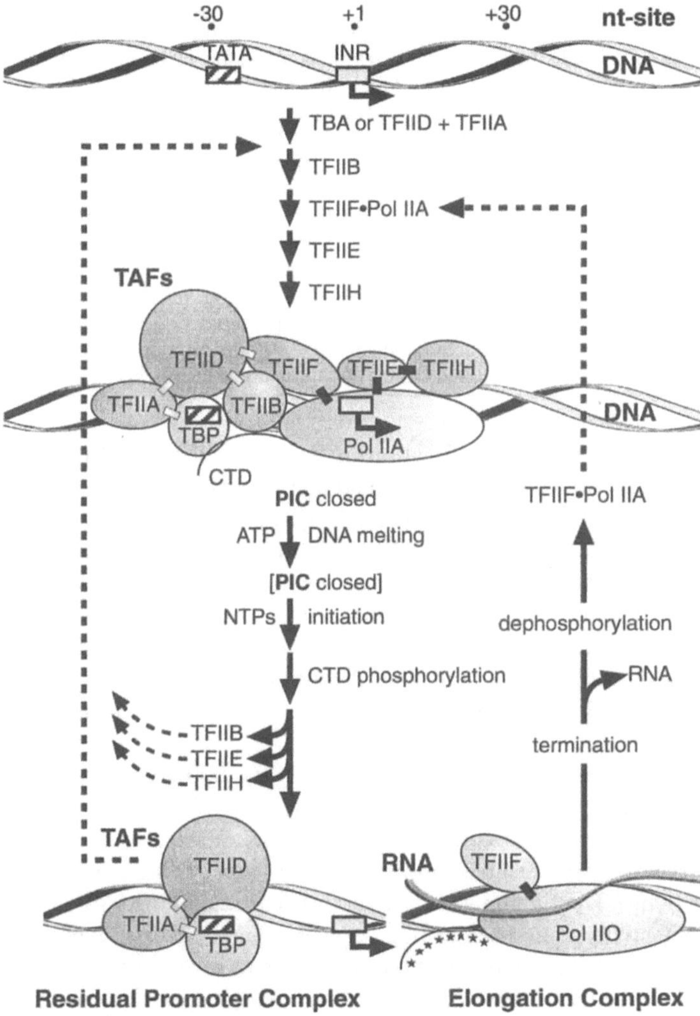

Fig. 5. Legend see p. 210

Fig. 5. *(see p. 209)* Signaling for transcription in nucleus of mammalian cells. Assembly of proteins in the formation of preinitiation complex (*PIC*) around a specific region of DNA begins with TFIID recognizing the TATA element, followed by coordinated assimilation of TFIIB, the phosphorylated form of RNA polymerase II (*Pol II*), and a complex array of transcription initiation factors: TFIIF, TFIIE, and TFIIH. If sufficient energy (ATP) is available to melt and open DNA chains and in the presence of sufficient nucleotide triphosphate (*NTP*) precursors, mRNA synthesis is initiated. During promoter clearance, the transcription factors TFIIB, TFIIE, and TFIIH dissociate from the TATA element, and an elongation factor forms from the phosphorylated form of Pol II and TFIIF. At the end of mRNA transcription, a phosphatase enzyme removes the phosphate group from Pol II and returns it to the original nonphosphorylated form to reinitiate the transcription process. Antisense oligomer blocking may be most effective in reducing the level of critical protein target below threshold concentration. (Modified from Paticoglou and Burley1997)

Oncoproteins mimic the function of growth factor, growth factor receptor at the cell membrane, GTP-binding protein (ras), mitogen-activated protein kinases (MAPK), and nuclear transcription factors (myc), thus promoting cell proliferation and giving the cancer cell its autocrine function (autokinase). Identification of the oncogenes in cancer cells and their mode of operation via the oncoproteins has provided us with a new tool with which to intervene in the process of protein cascades to inhibit cell proliferation (Gall and Pardue 1969; John et al. 1969; Zamecnik and Stephenson 1978; Southern 1975; Benton and Davis 1977).

We now know that approximately 70,000–100,000 genes (Table 1) are present in the human genome, consisting of 3×10^9 base pairs; only 4%–5% of these bases are used for making proteins. We also know that only a few diseases result from defects in a single gene, e.g., cystic fibrosis (lung), Huntington's disease (brain). The remaining 98% of the major potentially fatal diseases (e.g., cancer, atherosclerosis, hypertension) result from multiple defects or mutations in several genes. If we knew the sequence of these genes in normal cells, we would be able to identify susceptible populations by analyzing the genes of nucleated cells, e.g., lymphocytes in the blood. Cytogenetic studies have demonstrated several abnormalities and rearrangements of genes, e.g., translocations of a gene from one chromosome to another, deletions of bases resulting in the synthesis of defective proteins, and amplifications of certain genes (more than two copies of same gene in a normal cell).

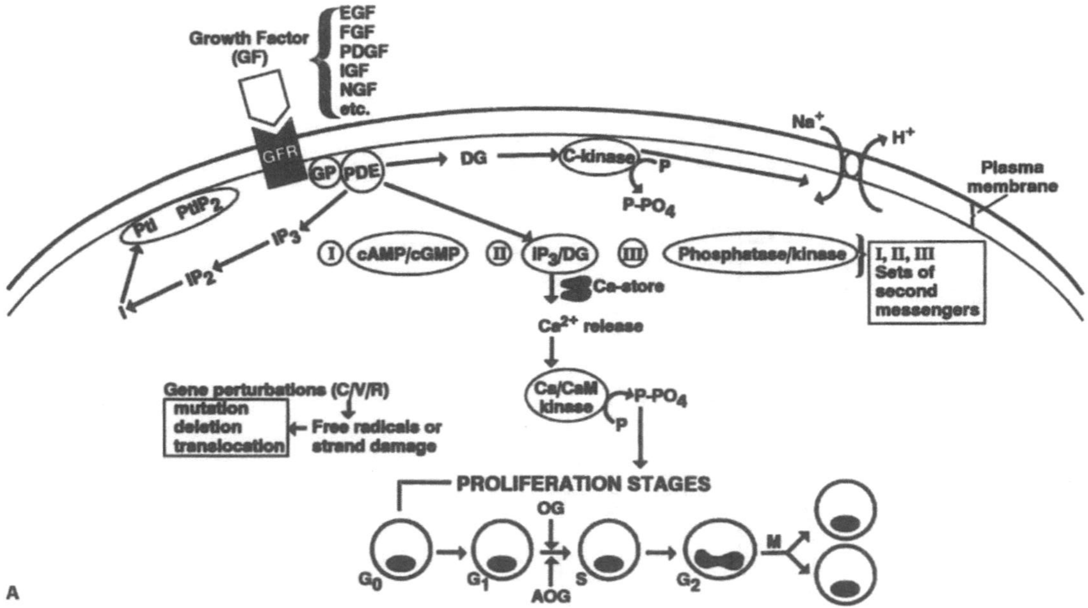

Fig. 6. A Regulation of transcription by growth factors. *GFR*, growth factor receptor; *PtI*, phosphatidyl inositol; *PtIp2*, phosphatidyl inositol diphosphate; *GTP*, guanidine triphosphate; *GP*, GTP-binding protein; *cAMP*, cyclic adenosine monophosphate; *cGMP*, cyclic guanidine monophosphate; *P*, protein, *P-PO4*, protein phosphate; *C/V/R*, chemical, virus, or radiation; *D*, differentiation; *UD*, undifferentiation; *M*, mitotic phase (prophase, metaphase, anaphase, and telophase); *OG*, oncogenes; *AOG*, antioncogenes. Proliferation stages: G_0, resting phase; G_1, gap 1; *S*, DNA synthesis; G_2, gap 2; *M*, mitosis

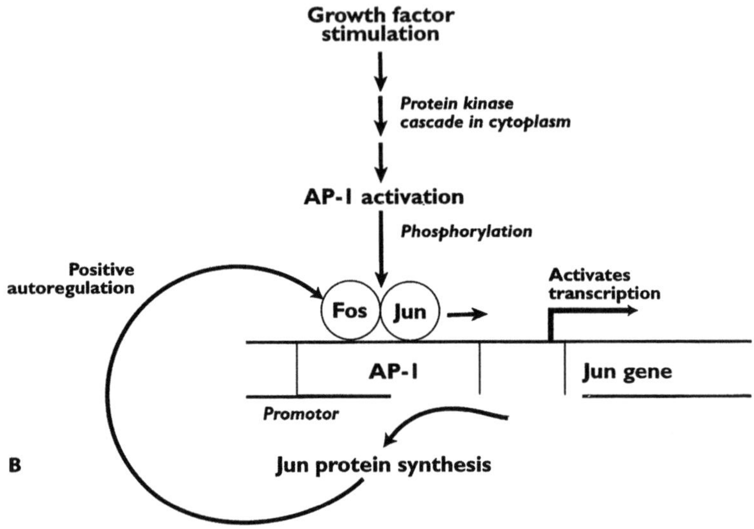

Fig. 6. B Regulation of transcription by c-fos-jun transcription factor

---

Fig. 7. Stress (cytokine)-induced signaling pathway in inflammation. *NF*, nuclear factor; *PKC*, protein kinase C ▶

Fig. 8. Regulation of cell cycle by the tumor suppressor gene Rb

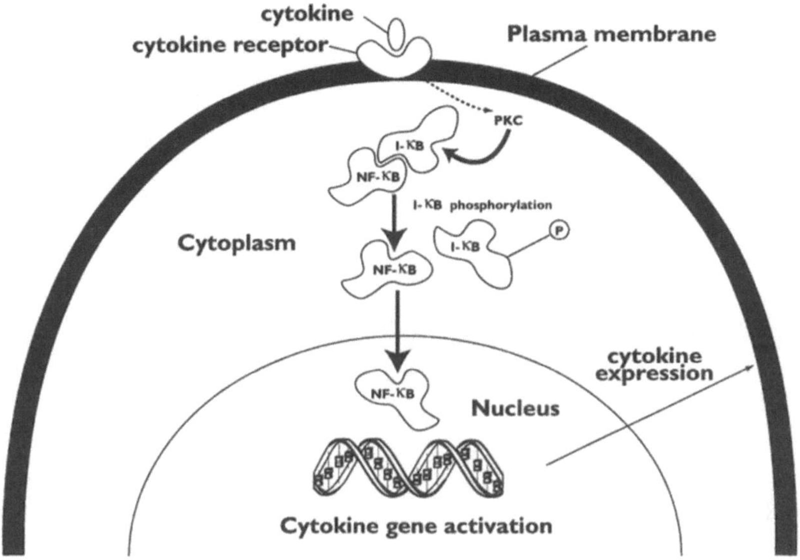

Fig. 7. Legend see p. 212

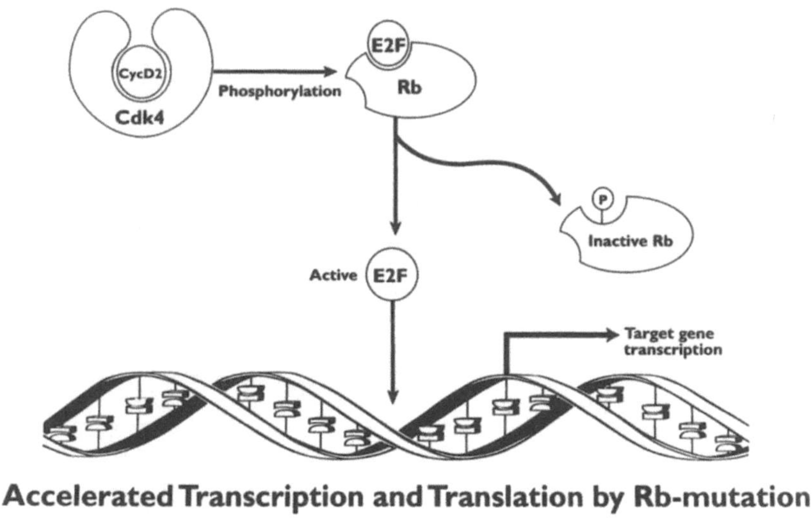

Fig. 8. Legend see p. 212

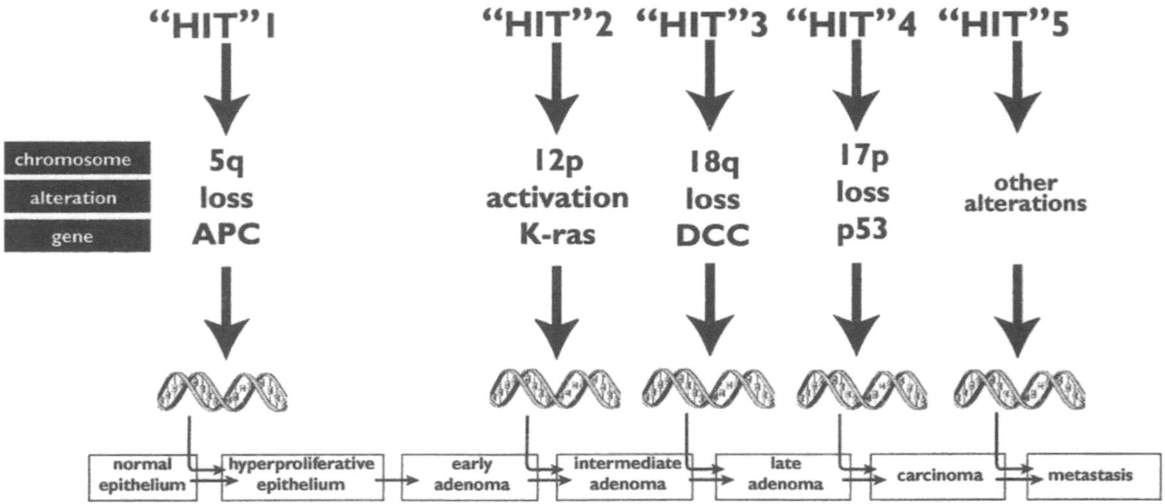

Fig. 9. Multiple-hit theory of colon cancer induction: activation of oncogenes and simultaneous loss of tumor suppressor genes. Acccumulation of genetic changes occurring over decades induces all forms of cancer. Inheritance of a single altered gene can result in a marked predisposition to colorectal cancer in two syndromes: familial adenomatous polyposis (FAP) and hereditary nonpolyposis colorectal cancer (HNPCC). A genetic defect in FAP affects the gatekeeper function of APC gene and initiates tumor. HNPCC induces tumor progression by targeting the genome guardian function of DNA mismatch repair. (Modified from Kinzler and Vogelstein 1996)

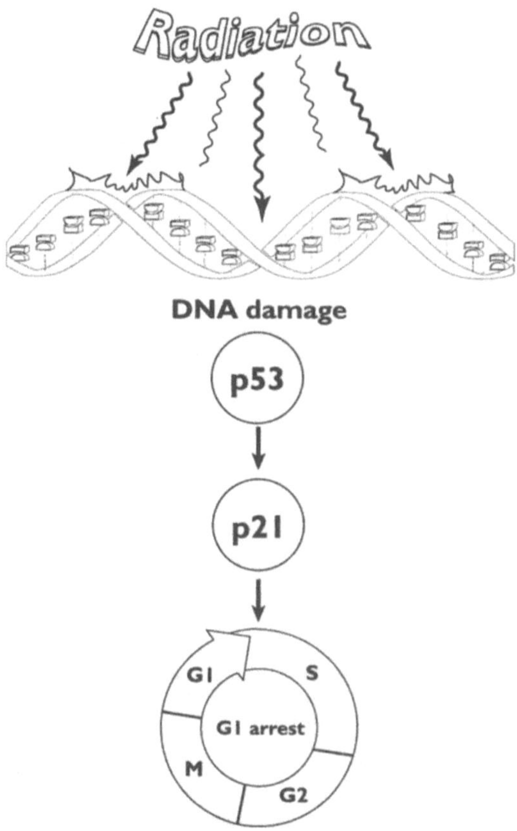

Fig. 10. Inhibition of cyclin-dependent kinase (CDK) by p53 and p21 after DNA damage by radiation and arrest of cell cycle in G_1 phase. Radiation of damaged vessel wall with γ-ray with a range of 10–12 mm after angioplasty along with antisense oligonucleotide treatment may retard the high rate of restenosis by migration from vessel media to intima and intimal proliferation of smooth muscle cells

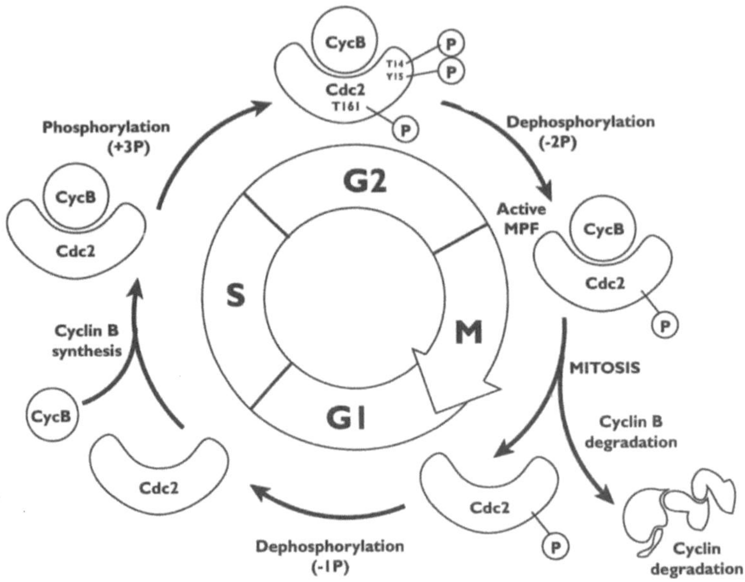

Fig. 11. Signaling for cell cycle regulation and DNA replication. DNA replication depends on activation of specific proteins at the checkpoint and hence may be assisted by accelerated transcription. *MPF*, maturation promoting factor (Modified from Alberts et al. 1994)

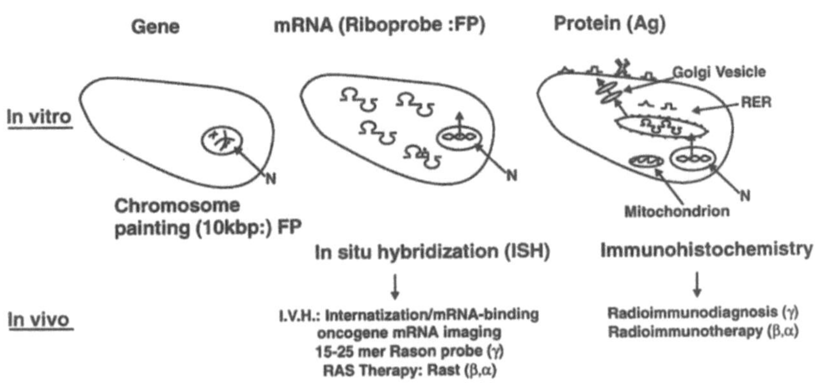

Fig. 12. Legend see p. 217

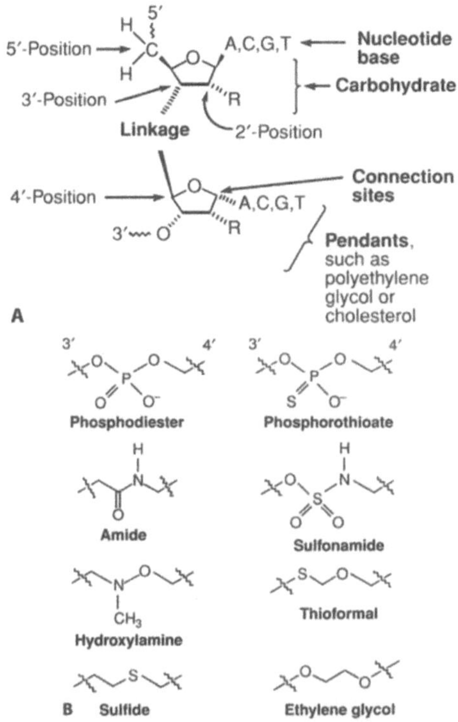

Fig. 13. A Antisense variations incorporate oligonucleotide modifications at base of nucleotides, carbohydrate moiety, and original phosphodiester linkages (From Baum 1994). **B** Phosphodiester linkages and other deoxyoligonucleotide derivatives with modified phosphate-sugar backbone. In the methylphosphonate (P-CH_3), phosphorothioate (P=S), and phosphoroamidate (P-N) internucleotide linkages, a single nonbridging oxygen atom is replaced by a methyl group, sulfur atom, or substituted amine. Although the derivatives are nuclease resistant and hybridize efficiently with complementary nucleotide sequence, the degradation product of mononucleotide phosphorothioate is also incorporated into genomic DNA. In addition, the P center is rendered chiral by these modifications. Achiral P analogues or stereoselective oligonucleotides are evaluated for higher potency. (Courtesy of Chem & Eng News, April 1994)

◀ **Fig. 12.** Detection of altered DNA, mRNA, and protein in in vitro and in vivo systems with protein- and nucleic acid-based diagnostic agents. Only tumor antigens and mRNA can be targeted in vivo systems. RER, rough endoplasmic reticulum

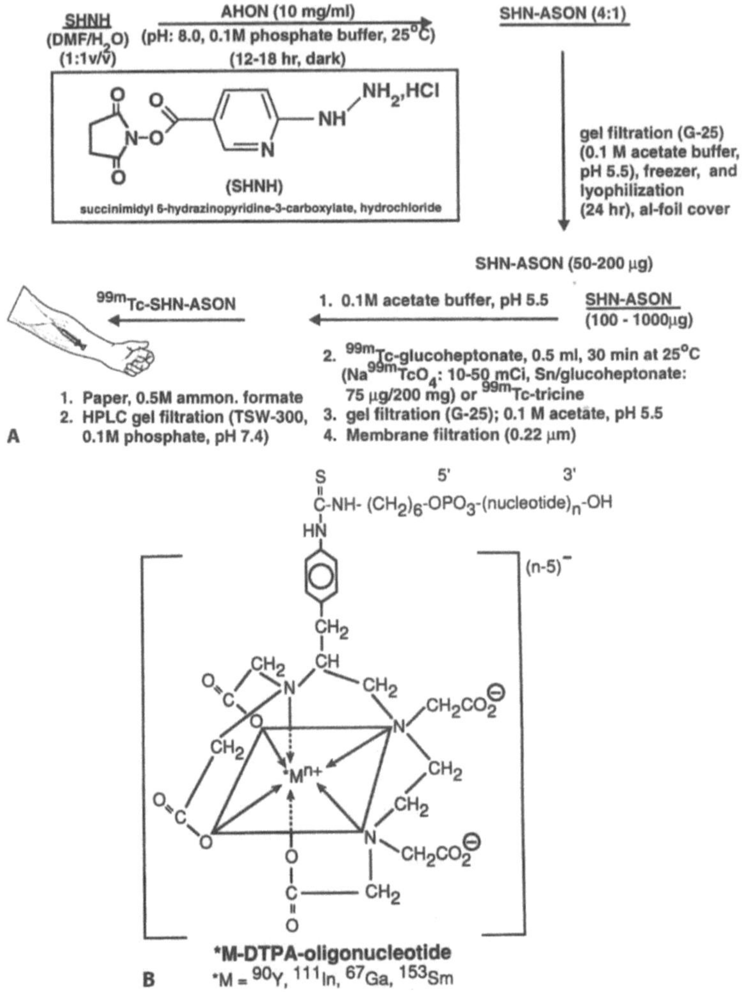

Fig. 14. A Conjugation of antisense deoxyoligonucleotide with succinimidyl hydrazinonicotinate (*SHN*) and kit formulations for labeling at highly specific activity with Tc-99m radionuclide. *HPLC*, high-performance liquid chromatography. *ASON*, antisense oligonucleotide. **B** General structure of oligonucleotide labeled with trivalent radionuclides. *DPTA*, diethylenetriaminopentoacetic acid

Table 1. Diversity of choice of target molecules besides receptors and enzymes for development of drugs and tracers in human genome: multiple genes, messenger RNAs, and proteins as potential targets for conventional drug and gene therapy

Genome size(base pair)	$3.5\times(10^9)$, 95% junk gene
Genes	100000 (100000–1000000 bp)
Oncogenes	100
Tumor suppressor genes	8–12
Messenger RNAs	100000 (500–8000 bases)
Proteins (structure and function)	100000
Differentiated cells	250
Total cells (Homo sapiens)	70×10^{12}

Table 2. Multiple interactions in the cellular signaling pathways, their sites of action, and relative affinities: master target paradigm

Interaction	Cell site	Affinity constant (l/mol)	Melting temperature
Protein–protein	Membrane	10^{-6}–10^{-12}	
(Peptide–protein)	Cytoplasm	10^{-6}–10^{-12}	
(Antigen–antibody)	Nucleus	10^{-6}–10^{-12}	
Protein–DNA	Nucleus	10^{-6}–10^{-10}	
Protein–mRNA	Cytoplasm	10^{-6}–10^{-10}	
	Nucleus	10–6–10^{-12}	
Ligand–receptor	Membrane	10^{-6}–10^{-12}	
ASON–mRNA	Cytoplasm		>70°C
ASON–DNA	Nucleus		>70°C
Chimeric ASON–mRNA	Cytoplasm		>72°C
Ribozyme–mRNA	Cytoplasm		>65°C
ON–protein (Aptamer, G quartet, small molecule, and macromolecule)	External cell matrix	10^{-10}	

For nucleic acid hybrid, higher affinity is measured in terms of higher melting temperature.
ASON, antisense oligonucleotide and its conjugated derivatives.

Table 3. Signal blockade in several diseases by antisense oligonucleotides in animal models

Target	Route	Species	References
Cardiovascular models			
c-*myb*	Topical	Rat	Simons et al. 1992
c-*myc*	Topical	Rat	Bennett et al. 1994
cdc2 kinase	Topical	Rat	Morishita et al. 1993 Abe et al. 1994
PCNA	Topical	Rat	Morishita et al. 1993 Simons et al. 1994
CDK2	Topical	Rat	Abe et al. 1994
Cyclin B_1	Topical	Rat	Morishita et al. 1994
E_2F	Soaking vein before grafting	Human saphenous vein bypass graft	
Angiotensin-1 receptor	Intracerebral	Rat	Gyurko et al. 1993 Phillips 1994
c-fos	Intracerebral	Rat	Suzuki et al. 1994
Inflammatory models			
Type I IL-I receptor	Intradermal	Mouse	Burch 1991
ICAM-I	Intravenous	Mouse	Stepkowski et al. 1994, 1995
Cancer models			
N-*myc*	Subcutaneous	Mouse	Burgess et al. 1995 Whitsell et al. 1991
NF-κB p65	Intraperitoneal	Mouse	Kitajima et al. 1992
NF-κB p65	Subcutaneous	Mouse	Higgins et al. 1993
c-*myb*	Subcutaneous	Mouse	Hijiya et al. 1994 Ratajczak et al. 1992
Ha-*ras*	Intratumor	Mouse	Schwab et al. 1994
BCR-ABL	Intravenous	Mouse	Skorski et al. 1994
Neurological models			
c-*fos*	Intracerebral	Rat	Chiasson et al. 1992 Heilig et al. 1993 Hooper et al. 1994 Woodburn et al. 1994
c-*fos*	Intraspinal	Rat	Gillardon et al. 1994
c-*jun*	Intracerebral	Rat	Tischmeyer et al. 1994
CREB	Intracerebral	Rat	Konradi et al. 1994
SNAP-25	Intracerebral	Chicken	Osen-Sand et al. 1993
Kinesin heavy chain	Intravitreal	Rabbit	Amaratunga et al. 1993
Tyrosine hydroxylase	Intracerebral	Rat	Skutella et al. 1994
GAD65	Intracerebral	Rat	McCarthy et al. 1994
GAD67	Intracerebral	Rat	McCarthy et al. 1994
Arginine vasopressin	Intracerebral	Rat	Flanagan et al. 1993
Insulin growth factor I	Intracerebral	Rat	Castro-Alamancos et al. 1994

Table 3. Continued

Target	Route	Species	References
Progesterone receptor	Intracerebral	Rat	Mani et al. 1994 Pollio et al. 1993
Estrogen receptor	Intracerebral	Rat	McCarthy et al. 1993
Oxytocin	Intracerebral	Rat	Neumann et al. 1994
Dopamine D_2 receptor	Intracerebral	Mouse	Zhou et al. 1994 Ogo et al. 1994
Dopamine D_1 receptor	Intracerebral	Mouse	Zhang et al. 1994
Substance P receptor	Intracerebral	Mouse	Ogo et al. 1994
Neuropeptide Y	Intracerebral	Rat	Akabaytashi et al. 1994
Y-Y1 receptor	Intracerebral	Rat	Wahlestedt et al. 1993
K-Opioid receptor	Intracerebral	Rat	Adams et al. 1994
δ-Opioid receptor	Intracerebral	Mouse	Lai et al. 1994
NMDA receptor	Intracerebral	Rat	Kindy 1994
CCK_R receptor	Intracerebral	Mouse	Vanderah et al. 1994
Virus infection models			
Duck hepatitis	Intravenous	Duck	Offensperger et al. 1993
HSV-1	Mouse		Kulka et al. 1989

Modified from Bennett and Crooke (1996).
PCNA, proliferating cell nuclear antigen; IL, interleukin; ICAM, intercellular adhesion molecule; NF, nuclear factor; CREB, cAMP response element-binding protein; NMDA, *N*-methyl-d-aspartate; CCK, cholecystokinin; HSV, herpes simplex virus; BCR-ABL, breakpoint cluster region-Abelson leukemia.

Table 4. Nonspecific effects of antisense oligonucleotides

Type of activity	Sequence specific
Inhibition of thrombin	Yes
Inhibition of gp120–CD4 interaction	Yes
Binding to CD4	No
Binding to bFGF	No
Inhibition of binding of interferon-γ receptor	Yes
Inhibition of PDGF binding to PDGF receptor	No
Inhibition of type II phospholipase A_2	Yes
Binding to NF-κB	No
Induction of Sp1 binding activity	No
Induction of protein kinase C activity	No
Inhibition of DNA polymerases	No
Inhibition of RNase H	No
B cell mitogen	Yes
Inhibition of plasmodium proliferation	No
Inhibition of epithelial cell proliferation	Yes
Inhibition of smooth muscle cell proliferation (reocclusion postangioplasty)	Yes
Inhibition of smooth muscle cell proliferation (saphenous vein bypass graft)	Yes
Inhibition of herpes simplex virus growth	Dose dependent
Inhibition of cell adhesion to substratum	Yes
Inhibition of in vitro translation reactions	No
Inhibition of transferrin receptor expression	No

Modified from Bennett and Crooke (1996).
bFGF, basic fibroblast growth factor; PDGF, platelet-derived growth factor; NF, nuclear factor.

Table 5. Advantages and disadvantages of antisense oligonucleotide peptide-based radiopharmaceuticals and radioimmunoconjugates

	ASON-RP	*Peptide-RP*	Radioimmunoconjugate
Size (Da)	4000–5000	4000–7000	160000, 100000, 50000
Permeability	High	High	Low
Half-life	1–2 h	1–2 h	24–48, 24, 6–8 h
Virus contamination	None	None	Human and murine origin (10%–15%)
Synthesis/production	Easy	Easy	Costly/laborious
RL tolerance[a]	Moderate	Moderate	High
Immune response	Low	None to low	HAMA (5%–30%)
Cross-reactivity	None	Low	Low
Adverse reactions	None	None to low	Fever, chill, rash, shock

ASON-RP, antisense oglionucleotide-regulating protein; HAMA, human anti-mouse antibody.
[a]Tolerance of ligand for radiolabeling without losing receptor-binding affinity.

By developing radioactive and colored fluorescent probes, we would also be able to detect these defective segments on glass slides and test tubes. A recent development is the γ-emitting radioactive probe. These probes are injected intravenously and subsequently search for cancer cells in the body; diagnosis of these cytogenetic abnormalities is made possible with the help of penetrating γ-radiation from cancer sites inside the organs using a γ-camera as the external detector. This has enabled these sensitive tools of molecular biology to be brought the patient's bedside for both diagnosis and therapy (Grunstein and Hogness 1975; Benton and Davis 1977; Sambrook et al. 1989).

12.2 Roles of Cytokines in Treatment of Inflammation, Infection, and Septic Shock

Although most of the antisense therapies were directed and designed for cancer and viral diseases, recently, new targets in the signaling pathways of inflammation have also been evaluated in animal models and patients. Investigators in biotechnology and drug companies have been working on the mechanism of inhibiting the inflammatory properties of cytokines, specifically tumor necrosis factor (TNF), for treatment of septic shock, autoimmune diseases (multiple sclerosis), rheumatic arthritis, and Crohn's disease (Fig. 7). TNF could then be used directly for killing tumor cells, something which is currently impossible due to overstimulation of the inflammatory response. Several proteins, e.g., TNF receptors (TNF-R1 and TNF-R2) and death-domain proteins, trigger the activation of NF-κB and promote inflammatory response and apoptosis. Several signal cascade proteins (Wallach 1997), i.e., MACH1 and MORT1, recognise Fas/APO1, inducing cell death. Metalloproteinases cleave the transmembrane form of TNF (26 kDa). After cleavage at the Ala-76–Val-77 bond, the mature form of TNF (17 kDa) is then released in plasma. The metalloproteinase involved is called TNFα-converting enzyme (TACE). The anti-inflammatory cytokine interleukin (IL)-10 inhibits inflammation by T lymphocytes via IL-1 and TNF. Intestinal ulceration in patients with Crohn's disease was significantly decreased by the injection of anti-TNF antibody. SmithKline Beecham Pharmaceuticals Inc. observed that a pyridinyl imidazole drug (SB203580) selectively inhibits the stress-activated protein kinase SAPK-2, an enzyme that is highly activated by bacterial endotoxins, lipopolysaccharide, and cytokines. Prompt selective blocking of SAPK may be more beneficial in blocking the downstream activation and amplification of inflammatory molecules rather than scavenging the plasma TNF by the TNF antibody (Briscoe et al. 1996; Maltese et al. 1995; Narayanan et al. 1993; Stepkowski et al. 1995).

12.3 Noninvasive Diagnostic Imaging with Radiopharmaceuticals (Tracers)

In nuclear medicine, several types of tracers are used to locate functioning and nonfunctioning (i.e., infected, cancerous, infarcted/dead or oxygen-deficient/ischemic) tissues of different organs. Tracers were designed to take advantage of changes of metabolic rate and other normal/abnormal functions in normal and abnormal tissues. Higher concentrations of tracers in diseased tissues make it easy to distinguish this tissue from normal tissue using radiation contrast generated on X-ray film or other forms of storage disk for processed image display. In this review, the technology of locating oncogenes via their mRNA product in cancer cells in live animals using radiolabeled antisense oligonucleotide (ASON) probes is described (Fig. 14). Until now, oncogenes in tumor tissue could only be tested by biopsy specimen, but by radiolabeling the oligonucleotide probes with γ-emitting radionuclides, we were able to localize tumor at an early stage in tumor-bearing mice using noninvasive imaging with a γ-camera, which identifies the higher amount of radiation from cancerous tissue. Cancer in humans might possibly be able to be diagnosed in the same way in the near future.

In this chapter, we focus on mRNA target and its in vitro and in vivo assay (Figs. 1, 2); we also discuss how to make radiolabeled, single-stranded short DNA probes to measure the mRNA of oncogenes and how this microscopic approach of cancer cell detection can be extended from test tubes or glass slides to living animal models.

RNA is analyzed by northern blotting; it is single stranded and less stable than DNA, but follows the same rule of hybridization by hydrogen bonding of complementary bases on opposite strands (A with U and G with C). The blotting protocol for DNA (transfer from soft, difficult-to-handle gel to tough, plastic membrane) was originally developed by Southern in 1975 and was subsequently extended to mRNA (northern blotting) and protein (western blotting). RNAs isolated from cells are denatured to straighten them and then separated according to size by electrophoresis; they are transferred from soft agar gel by placing a stack of paper on the gel and membrane to promote transfer (capillary action in cellulose fibers) of mRNA by appropriate buffer to sturdy nitrocellulose membrane. With a replica of mRNA distribution in gel, this membrane is baked to immobilize mRNA. The membrane is hybridized with radiolabeled mRNA probes to detect the mRNAs of oncogenes or other genes of interest in cancer cells and tissues. Conventional mammography occasionally misses the site of breast cancer that lies hidden in dense, calcified breast tissue and appears as dense cloud. Biopsy with a fine needle or excision may be necessary to confirm the presence of a breast cancer nodule.

Radiochemists, nuclear chemists, and medicinal chemists have been designing new radiolabeled probes for the early diagnosis of cancer. Using radiolabeled antibodies, we have been able to detect metastatic sites when the tumor mass is 2–3 cm in size. The "magic bullet" of labeled mouse antibody is large and there are many diffusion barriers before reaching the target antigens in tumor. The principle of this test (radioimmunoassay or radioimmunodetection) depends on the binding of specific amino acid sequence of radiolabeled antibody with that of epitope of the tumor-specific protein, called tumor-associated antigen. Several radiolabeled high-affinity peptides are also being tested for noninvasive diagnosis of infection, site of arterial plaque (atheroma), and tumor (Dewanjee and Serafini 1996).

12.3.1 Nucleic Acid-Based Radiolabeled Probes

The nucleic acids of DNA and RNA are characterized by their size (length or number of base pairs for DNA and bases for mRNA) and nucleotide sequence. Size is determined by electrophoresis; acids with a shorter length migrate over a longer distance in the denaturing media. Similarity of nucleic acid segments, their size, and their amount are determined by labeled probes. The probe is synthesized chemically or enzymatically so that it has a nucleic acid sequence that hybridizes with all or part of the target DNA or RNA and a marker of radiation, fluorescence, or chemiluminescence. DNA and RNA probes labeled with ^{32}P or ^{35}S (substitution for oxygen in the phosphate group) are detected by autoradiography with sensitive X-ray film. Nonisotopic markers strive to reach the level of sensitivity of radioactive probes. These markers bind to the chemical side group of the base of the nucleotide substrate, and the side group is ultimately recognized by a specific antibody or a binding protein. Commonly used tags include digoxigenin, a plant steroid, and biotin, a member of the vitamin B complex. The former is detected by an antibody and the latter by a high-affinity binding protein, avidin. The complex is measured by colorimetric or fluorescent assay. The advantages and disadvantages of three different types of probes are shown in Table 5.

DNA probes are prepared by nicking (cutting open) and labeling at the nicked site or by labeling at random sites. The former is called nick-translation labeling, which labels both strands of DNA, and the latter, random primer labeling. The nicks are made by DNase I (DNA-cutting enzyme) in both strands. Since the strands remain associated by multiple H-bonds, they can be repaired by incorporating radioactive substrates at the nicks with the help of DNA polymerase I (an enzyme that creates DNA polymer from single nucleotides) and DNA ligase (an enzyme that joins DNA segments). Before use, the two radioactive strands are separated by boiling so that they can be used as a single-stranded probe.

In random primer labeling, the strands are separated by boiling and single strands of oligonucleotide (6-mer) are added to the reaction mixture. These hybridize at various sites with each DNA strand, acting as a primer for the initiation of DNA synthesis by DNA polymerase I by incorporating radioactive nucleotide substrates from the reaction mixture. The major disadvantage of this method is its lower sensitivity due to the ability of the primers to hybridize with natural complement, thus losing their ability to hybridize with the intended strands. Single-stranded complementary RNA (cRNA) probe is a better alternative amd is made by in vitro transcription. Bacterial plasmids are available that contain RNA polymerase promoter region adjacent to a restriction site for inserting cDNA. The directional promoter guides the synthesis of inserted sequence of intended single strands in the plasmid (circular DNA found in bacteria).

12.3.2 Hybridization of Radiolabeled Oligonucleotide Probe with mRNA

In situ hybridization (ISH) is a logical extension of the standard procedures of molecular biology, i.e., radiolabeling of nucleic acid probes and their applications for detection of mRNA and DNA in specific locations in the cytoplasm, organelles, or nucleus. ISH procedures used radiolabeled cDNA, riboprobe, or oligonucleotide probe. Most of these probes are labeled with β-emitting radionuclides of tritium (^{3}H), phosphorus (^{32}P, ^{33}P), or sulfur (^{35}S). These synthetic oligonucleotides freed us from the shackles of enzymes and labor-intensive biochemical procedures. As the length of nucleotides increases, so does the amount of failure sequences with a shorter length. These sequences must be removed by molecular sieving techniques before conjugation and radiolabeling procedures are performed. Enzymatic methods are essential for longer DNA and riboprobes.

Using radiolabeled oligonucleotides, we extended this approach one step further to visualize specific mRNAs of cancer cells. Most cancer cells contain mutated proto-oncogenes, known as oncogenes, which have been isolated from many retroviruses (RNA tumor viruses) and a few tumors. During their evolution, these viruses captured genes from normal cells that gave them powerful growth-transforming properties. Rous (1911) was the first to discover the Rous sarcoma virus (RSV), which possesses a specific gene site termed "src." src-related oncogene sequences were found in their natural host (chicken) and in other vertebrates, including humans. Using cloning techniques, we were able to make large amounts of these mutated genes and to study the roles associated with their ability to transform normal cells into cancer cells. Inoculation of these retroviral gene-containing cells in normal animals causes rapid tumor growth. The viral form of such oncogenes is labeled with the prefix "v" (e.g., v-*src*), and the corresponding cellular gene is labeled "c" (e.g., c-*src*).

We focused our investigations on amplified oncogenes. In leukemia and cancer of some of the organs, multiple copies of several oncogenes, e.g., c-*myc* and c-*erb*B2, are present. They are found to make many more copies of mRNA than normal cells. Binding of radiolabeled probes with many mRNA copies in cancer cells and few copies in normal cells results in higher contrast, making it easier to visualize the cancer cells among normal cells.

12.3.3 Rational Development of Nucleic-Acid Based Radiopharmaceuticals

Several oncogenes have been identified in human solid tumors, and we now know the sequence of nucleotides in these oncogenes. Specific small, single-stranded DNA segments can be made by passing along this strand and picking the correct sequence by trial and error using some empirical equations. Access to DNA and mRNA is limited, since DNA is coiled around the compact balls of basic nucleoproteins called histones and mRNA is associated with other proteins. Small, single-stranded DNA segments, called oligonucleotides, bind (hybridize) the specific nucleotide sequence of oncogene mRNA in cancer cells (ISH) and help to locate the primary and metastatic sites in animal models; since oncogenes are conserved in several species of animals, we believe that these probes or their analogues might also be used in cancer detection in humans. Currently, biopsy specimen is necessary to identify these oncogenes and cytogenetic abnormalities. High concentrations of these oncogenes in specimen suggest poor outcome and a short survival time for the cancer patient. The penetrating radiation of γ-rays emanating from cancer sites in patients could be detected using a γ-camera.

Among the cytogenetic changes observed in cancer cells, the DNA segments appearing as homogeneously staining regions (HSR) occur in chromosomes due to multiple copies of mutated or translocated genes (moved from one chromosome to another). Using colored DNA probes, these areas could be identified in spread chromosomes of cancer cells by a process called chromosome painting. In some cases, amplified DNAs are also present as minute double chromosomes; these small chromosome-like structures lack centromeres and are not consistently inherited at cell division. When integrated in chromosomes, they appear as HSR. Since these multiple copies of genes in a single cancer cell make more mRNAs and more proteins, these markers are suitable targets for location using a radioactive DNA probe and a radioactive antibody. Two such oncogenes are *myc* and *erb*B2, which are found in cancer cells in the blood and in other solid tumors of brain, lung, breast, colon, and lymph nodes (Watt et al. 1983; Weinberg 1996).

ASON have been found to inhibit a variety of oncogenes (c-*ras*, c-*myc*, c-*myb*, bcr-*abl*). By binding with a specific segment of mRNA, they prevent synthesis of new proteins. Phosphorothioates and methylphosphonates were found to be nuclease resistant and more effective in inducing phenotypic changes and inhibition of gene expression at high concentrations (10–100 m*M*). This inhibition may result from the antisense and nonspecific effects of a polyanion. The sequences working on the antisense mechanism were chosen for generation of antisense radiopharmaceuticals (Dewanjee et al. 1991; Dewanjee and Serafini 1996).

12.3.4 Mechanism of Inhibition by Antisense Oligonucleotide

After probe internalization, the mRNA target in the cytoplasm is bound by the probe. Oligonucleotide can inhibit gene expression by the following three mechanisms:

1. Formation of RNA–ASON duplex followed by cleavage of the RNA portion by RNaseH; the cleaved and incompetent RNA is degraded, thus inducing translational arrest in the target cell.
2. RNA–ASON duplex may inhibit gene expression by steric hindrance blocking the interaction of mRNA with other cellular components, thus preventing specific protein sysnthesis. In the cytoplasm, steric blocking may inhibit translation initiation and ribosome elongation. ASON binding to areas of base pair mismatches may generate short-lived complexes without disrupting the expression of the nontarget genes.
3. Steric blockade in the nucleus may inhibit pre-RNA splicing, polyadenylation, or mRNA transport from the nucleus to the cytoplasm.

The concept of "stringency" of conditions of hybridization is very important in understanding the specificity of reactions between nucleic acid probes and target mRNA. The probes can tolerate a certain small number of mismatched base pairs (e.g., A with C or G with T) and still form stable duplexes; if the degree of mismatch is high and extracellular and cytoplasmic milieus are inappropriate, the probe–mRNA duplex or hybrid will fall apart and the probe cannot be used for detection. Favorable in vitro conditions with considerable mismatch (low stringency) include low temperature, low concentration of formamide, and high concentration of salt. Under conditions of high stringency (high temperature, low salt concentration), only exact matches of DNA will anneal, especially when the probe length is short. Although these parameters can be optimized in ISH, they are within a tight range in normal and diseased cells of live animals.

12.4 Synthesis, Conjugation, Radiolabeling, Purification, and Characterization of Aminohexyl-Conjugated Oligodeoxynucleotides

The phosphodiester backbone of antisense oligodeoxynucleotide may be modified for diagnosis of cancer and infection and for more effective therapy. Bearing in mind the nuclease sensitivity of the phosphodiester backbone, several modifications have been made to enhance stability, permeability, and target affinity. Most of these derivatives (two groups of phosphorus and nonphosphorus analogues) are nuclease resistant and may have chiral centers (CC) or activated ribonuclease H (RNaseH, RNHA; Gao et al. 1992). Occasionally, either alternating phosphodiester and novel linkages or nuclease-resistant derivatives are placed at the 3' and 5' ends. Several strategies are used to replace the phosphodiester backbone with novel linkages along with characteristics of nuclease resistance, chirality, and RNaseH activation (Figs. 13, 14),including the following:

1. Phosphorus analogues of phosphodiester linkages
2. Nonphosphorus analogues of phosphodiester linkages
3. Sugar modifications
4. Heterocyclic base modifications

The best radionuclides for diagnostic imaging are those of metallic elements. These are easily available and not very expensive and have plenty of γ-rays with suitable energy (80–250 keV). To bind these metallic radionuclides, the oligonucleotides are modified with a linker molecule which will attach a chelating molecule (polyaminopolycarboxylate, diethylenetriamine pentaacetate, DTPA; succinimidyl hydrazinonicotinate, SHN) by an amide/ester bridge on one end and a chelate radiometal on the other end (Fig. 14). The general structures of synthesized oligodeoxynucleotide, aminohexyl oligodeoxynucleotide, DTPA-coupled oligodeoxynucleotide, and the corresponding ^{111}In chelated probe are shown below:

- 5'-OPO_3-oligonucleotide-OH-3'
- 5'-NH_2-$(CH_2)_6$-OPO_3-oligonucleotide-OH-3'
- 5'-DTPA-CSNH-$(CH_2)_6$-OPO_3-oligonucleotide-OH-3'
- 5'-^{111}In-DTPA-CSNH-$(CH_2)_6$-OPO_3-oligonucleotide-OH-3'

The 15- to 25-mer oligonucleotides, carrying an aminohexyl group at their 5' ends, were synthesized on a DNA synthesizer using the dimethoxytrityl nucleoside phosphoramidite triester coupling method on a 10-m*M* solid support column and were amino-linked, deprotected, and purified in an ion-exchange high-performance liquid chromatography (HPLC) column. Tetraethylthiuram disulfide (TETD) in acetonitrile converts internucleotide phosphite to the phosphorothioate triester. Aliquots (0.1–1.0 mg) of purified deoxyoligonucleotide are best preserved by storing at –80°C. The primary amino group on these oligomers was conjugated with a benzylisothiocyanate derivative of DTPA or other analogues for labeling with ^{111}In or Tc-99m radionuclide. The conditions of conjugation of diethylenetriamine pentaacetate-isothiocyanate (DTPAITC) with aminohexyl oligodeoxynucleotide (DTPA-ASON) were evaluated and chelation of DTPAASON was optimized at a high stoichiometry of DTPAITC to ASON (10:1), a short incubation time of 30 min at room temperature, and a pH of 6.5. The free radionuclide was removed by molecular sieving (TSK-300 SW) with an HPLC system, and freeze-dried aliquots were stored at –25°C. The impurities in the labeled probes and their stability were checked with thin-layer and size-exclusion (HPLC) techniques (Dewanjee et al. 1993a, 1994a–c).

Four major issues have to be resolved: stability, permeability, radiolabeling at high specific activity, and radiolabeling with γ-emitting radionuclides without altering the properties of hybridization with mRNA. In addition, the nucleotide sequence of oncogenes needs to be determined. A compromise has to be made between permeability and specificity of hybridization. Although smaller probes are taken inside cells through the cell membrane, they may not attach strongly or may not be unique in binding the specific mRNA. Since the mRNA maintains a three-dimensional structure with hairpins and bulges, a base sequence has to be selected that will bind without a single mismatch of a single base. These probes are taken inside the cytoplasm (Loke et al. 1989) from outside the cell membrane via a specific shuttle protein carrier.

Preliminary work by other investigators also indicates that these small sequences of 15–25 nucleotides pass through the membrane of live cells and bind with mRNA, preventing translation of the specific message and hence the formation of the specific protein. We initially relied on these nucleotide sequences to establish our methodology and demonstrated that in vivo hybridization techniques are feasible after labeling with γ-emitting radionuclides of iodine (^{123}I), indium (^{111}In), and technetium (Tc-99m). For radioiodination, we needed an activated phenyl ring with the ability to conjugate via the isothiocyanate to form the thiourea linkage, which is more stable than amide linkage.

We have validated our in vitro and in vivo hybridization methods using cultured cell suspensions, tissue slices, and tumor-bearing mice. It was necessary to convince both ourselves and our critics, who are very concerned about enzymatic attack and breakdown of probes in plasma and tissues. These enzymes of endonucleases (cut in the middle of the sequence) and exonucleases (cut at the ends of sequence). We were surprised to find that the bulky chelating molecule with a six-carbon spacer that we attached to the nucleotide sequence, inhibited nuclease attack. Investigators at the University of Miami, School of Medicine Cancer Center (Dr. A. Krishan and Dr. D. Lopez), provided us with cancer cell lines and tumor-bearing mice to test these problems of stability, permeability, and binding with mRNA in cancer cells. A parallel sequence (sense) also has to be used, so that it can act as a control and stay in the cancer cells at a much reduced level. We are also using this approach to treat tumor cells, i.e., specific nucleotide sequence is being used to deliver low-, medium-, and high-energy β-emitting radionuclides (^{125}I, ^{89}Sr, ^{90}Y, ^{153}Sm) to kill tumor cells in live mice. For some of the β-emitters, the specific activity is low and cannot be effectively used in therapy. Further work is in progress to exploit the fruits of this laboratory work using mice for therapy of cancer patients (Dewanjee et al. 1994, 1995).

12.5 Design of Novel and Conventional Radiopharmaceuticals

Most rational research efforts in nuclear medicine in the last decade have focused on the development of radioactive ligands for the detection of a limited number of membrane antigens (tumor markers, somatostatin receptors on tumor cells) or receptors (e.g., muscarinic acetylcholine, dopamine, opiate), which are more readily accessible than cytosolic or nuclear receptors. However, these general concepts of radiolabeling methods using ^{123}I, ^{111}In, and Tc-99m have significantly helped us to embark on the next phase of development of γ-emitting, single-stranded DNA probes for studying gene activation (Dewanjee and Serafini 1996). During the last decade, significant progress has been made in the development of ASONs as interventional tools in pharmacological studies and therapeutic agents in several animal models of acquired immunodeficiency syndrome (AIDS), inflammation, and cancer, as well as for use as therapeutic agents in several clinical trials.

Using short, specific ASON sequence , complementary to mRNA sequence, selective inhibition of gene expression was demonstrated; when sense and scrambled sequences were used as control sequences, no such inhibition was observed. However, other effects not related to the antisense mechanism have also been obsrved in a number of in vitro cell cultures and in vivo studies in animal models. Previous studies also indicated the nuclease sensitivity of phosphodiester linkage (Wickstrom 1986); hence other derivatives, e.g., methylphosphonate and phosphorothioate, were developed. Further modifications of phosphodiester backbone, ribose, and nucleotide subunits identified other derivatives with improved potency, desirable pharmacokinetics, and reduced toxicity. Encouraged by in vitro and in vivo inhibition studies, we radiolabeled oligonucleotides with nonmetallic and metallic radionuclides of ^{123}I, ^{111}In, and Tc-99m and demonstrated their use in imaging specific mRNA of the oncogenes c-*myc* and *erb*B2, which are present in a mouse model of breast tumor. We also carried out noninvasive imaging of mRNA of heat shock protein (HSP-70) in a pig model during an ischemic episode of cardiopulmonary bypass. The potential cell-killing ability with ^{90}Y-labeled antisense probes was also observed in cultured cell lines.

12.6 Structure and Functions of Gene-Regulatory Proteins

Gene-regulatory proteins (GRPs) recognize the specific short stretches of double-helical DNA and activate specific genes to synthesize the corresponding mRNA and to express the protein. Several hundred GRPs have been identified in both prokaryotic and eukaryotic cells. These GRPs have unique structural motifs, e.g., helix–turn–helix motif, zinc finger motifs, leucine zipper motif, and helix–loop–helix motif. The folded amino acid sequences of these motifs bind DNA as a homodimer or heterodimer. These GRPs and their ingenious methods of gene activation were identified in bacteria and bacteriophage-λ about 40 years ago. The circular DNA of *Escherichia coli* (5×10^6 base pairs) encodes 4000 proteins. *E. coli* regulates the expression of these genes according to food sources available in the environment. Five *E. coli* genes encode enzymes to make the amino acid tryptophan. When there is enough tryptophan in the environment and in the cell, the gene is switched off.

Within the promoter that regulates the tryptophan genes lies the short segment of regulatory DNA called the tryptophan operator. It is recognized by a tryptophan repressor (a helix–turn–helix motif protein that binds the major groove of DNA) and occupies the promoter region, preventing access of RNA polymerase, when no tryptophan is needed. Five genes are transcribed as a single mRNA molecule in bacteria for making the enzymes necessary for tryptophan biosynthesis; this type of gene cluster is called an operon. The GRPs that function in this mode of negative control are called transcriptional repressors or gene repressor proteins. Some promoters are poorly recognized by RNA polymerase or are unable to open DNA for transcription. A group of proteins called transcriptional activators, transcription factors, or gene activator proteins bind to the neighboring site of promoter and RNA polymerase and dramatically increase the rate of transcription. Bacterial activator proteins, e.g., catabolite activator protein (CAP), can activate genes that enable *E. coli* to use alternative sources of carbon, when the preferred source, glucose, is not available. Lower glucose levels increase cAMP, which binds CAP and switches on genes of the *lac* operon to use the disaccharide lactose. The *lac* operon codes for proteins of lactose transport and its breakdown into glucose and galactose. This unique mode of gene control was recently explained by X-ray crystallography of DNA–protein complexes. The epitope involved in protein–DNA complex varies in length from four to 12 amino acids or more, and the recognition domain in DNA varies from four to eight nucleotides for restriction enzymes, with other genes and proteins having more nucleotides (Alberts et al. 1994; Lodish et al. 1995).

Figures 4 and 5 show how the general transcription factors assemble at the promoter to assist RNA polymerases (Pol II) in transcribing a new copy of mRNA. Eukaryotic transcription can be described in terms of preinitiation complex (PIC) assembly, PIC activation (DNA melting), initiation of transcription, promoter clearance, elongation, and termination steps. The assembly process starts with the binding of TFIID to the TATA sequence, a short DNA segment containing primarily T and A nucleotides. The TATA sequence is a component of nearly all the promoters used by Pol II and is typically located 25 nucleotides upstream from the transcription initiation site. TFIID is composed of many subunits called domains, and the TATA-binding domain is called TBP (TATA-binding protein). After TFIID binding, other transcription initiation factors, along with RNA Pol II, are added to assemble the functional executors. After tethering the Pol II to the promoter, it is freed of other transcription factors. For some promoters, phosphorylation of Pol II by TFIIH, a kinase, accelerates the disengagement process, allowing transcription to proceed. The site of phosphorylation is a long polypeptide tail that extends from the largest subunit of Pol II. In mammalian cells, Pol II is composed of 52 repeats of YSPTSPS including the phosphorylation sites of serine (S) and threonine (T). Another two eukaryotic RNA polymerases (Pol I, Pol III) require other sets of transcription factors, with the exception of the independence of TATA sequence of DNA. The general transcription factors are highly conserved from yeast to humans. Elucidation of the structural and functional features of these factors on model promoters has revealed insights into biochemical mechanisms for regulation on diverse promoters by gene and cell-specific activators. DNA sequences thousands of nucleotides away from the eukaryotic promoters can activate transcription at the promoter of gene; such sequences are known as enhancers. This activation or enhancement of transcription at a distance (50000 nucleotide pairs) may occur due to looping out of intermediary DNA sequence, thus permitting the proteins bound to promoter and enhancer to interact with the necessary polymerase (Patikoglu and Burley 1997; Roeder 1996; Kaiser and Meisterernst 1996; Bjorklund and Kim 1996).

12.7 Protein–Small Molecule, Protein–Protein, and Protein–DNA Interactions in the Signaling Pathways for Transcription, Translation, and Cell Cycle Progression

In the final phase of protein–DNA interaction for RNA transcription, RNA polymerase, transcription factor, and other cofactors have to bind the specific promoter region of DNA, the TATA-binding region (TBD). From the structural database of protein–DNA complexes furnished by X-ray crystallography and solution magnetic resonance imaging (MRI) spectroscopy, no specific code of nucleotide or amino acid sequence was found for the recognition of specific domains. Two universal features in these DNA–protein complexes are (1) rearrangement and loss of water molecules from both DNA and protein and (2) conformational changes of both DNA and protein on complex formation. In TBP, the interacting surfaces become dehydrated after complex formation. Borrowing ideas from protein folding for the analysis of protein–DNA recognition, Spolar and Record (1994) presented a thermodynamic picture of interactions in terms of hydrophobic effect (burial of nonpolar surface area, extrusion of hydrated water in the bulk solvent), loss of rigid body rotational and translational entropy, the polyelectrolyte effect of cation displacement on binding, and loss of conformational flexibility. The large heat capacity loss from DNA–protein formation originates from these entropic effects.

12.8 Apoptosis Program for Treatment of Cancer and Restenosis After Angioplasty

Ellis and Horvitz (1986) discovered that 131 cells in *Caenorhabditis elegans* (worm) larvae die by apoptosis. They observed that cell death (*ced*) genes that regulate apoptosis, including *ced*-3, encode a protease. The *ced*-9 gene was found to prevent apoptosis. Kerr et al. (1972) observed that, after treatment with glucocorticoids, the DNA of thymocytes is cleaved and forms a characteristic ladder pattern, identified in the electrophoregram of agarose gel. Korsmeyer (1995) discovered that Bcl-2 is a membrane protein that prevents apoptosis in B cell lymphoma and also observed that *bax*, which forms a *bax–bax* dimer, promotes apoptosis. Yonehara and Trauth et al. (1989) independently found that monoclonal antibodies they called anti-Fas and anti-Apo-1 could kill cultured human cells that bear Fas or Apo-1 antigen (Campuzano 1996; Gajewski and Thomson 1996; Livesey and Hunt 1996; Tomei and Cope 1991). The intracellular death domain of Fas has high homology with the *Drosophila* reaper protein. Several cysteine proteases, including calpain and ICE (interleukin-1β-converting enzyme), activate enzymes that participate in apoptosis. In addition, contradictory signals of growth induced by c-*myc* and growth arrest by p53 also trigger apoptosis.

12.9 Cell Signaling for Transcription, Translation, and DNA Replication

Cell signaling for transcription, translation, and DNA replication depends on multiple phosphorylations of downstream effector proteins via kinases involved in the signaling cascade process; approximately four to eight phosphorylations may occur in the vicinity of membrane, eight to 12 in the cytoplasm, and eight to 12 in the cell nucleus (Figs. 5–8). For the effector proteins to be recycled for the next round of phosphorylation, they have to be dephosphorylated by phosphatases. Efficient transcription, translation, and DNA replication involves minimal steps in signaling and the use of ATP molecules in a minimal amount of time, as in cancer cells. Mammalian cells communicate by means of hundreds of varieties of signaling molecules, e.g., peptides, proteins, amino acids, nucleotides, steroids, retinoids, fatty acid derivatives, and small molecules of diffusible and dissolved gases of nitric oxide and carbon dioxide. Most of the signaling molecules are secreted by signaling cells by exocytosis, while others diffuse through the plasma membrane. Some others are anchored to plasma membrane and influence only the contacting target cell. Proteins that drive the cell cycle, e.g., cyclins, cyclin-dependent kinases, and other accessory factors also have to be transcribed for DNA duplication. However, basal levels of some of these proteins are present in a variety of cells. In unicellular bacteria, a two-component system, a receptor tyrosine kinase and a transcription factor, is sufficient for signaling of transcriptional regulation; however, complexity and redundancy increase with prokaryotes, yeast, and multicellular organisms (Alberts et al. 1994; Lodish et al. 1995).

Most of the extracellular signaling molecules, e.g., growth factor, act on the receptor of target cells at low concentrations ($<10^{-8}$ *M*), and receptors bind them with high affinity ($K_a > 10^8$ l/mol). Most of the receptors are transmembrane proteins on the target cell surface. When they bind an extracellular signaling molecule (a ligand), the receptor becomes activated, thus generating a casacde of intracellular signals and altering the behavior of the target cell. For intracellular targets, the signaling ligand must be small and hydrophobic in order to diffuse across the plasma membrane (Figs. 1, 2). For some steps, e.g., transcription, several proteins with their epitopes exposed to the corresponding contact points are involved in the initiation of mRNA synthesis. In these complex interactions, the cumulative affinity constants of multiple interactions from DNA to proteins can be expressed in terms of cooperativity. Signaling by secreted molecules may act on distant targets or on neighboring molecules. In *paracrine signaling*, the molecules are not allowed to diffuse too far; they are taken up by target cells, destroyed by extracellular enzymes, or immobilized by extracellular matrix. The neurons contact their target at the end of axons known as chemical synapses, and the neurotransmitter is delivered specifically to the postsynaptic target. In *endocrine signaling*, the secreted hormone released in the bloodstream controls the behavior of the whole organism by means of a relatively slow process. In *autocrine signaling*, the secreted molecule binds back to its own receptor. This signal is effective on a specialized group of differentiated cells during development (Fig. 1A). Eicosanoids derived from fatty acids (arachidonic acid) cleaved from plasma membrane by phospholipase have a variety of effects associated with contraction of smooth muscle cells, aggregation of platelets by thromboxane, and pain and inflammation caused by leukotrienes. Signaling to neighboring cells is carried out by small molecules (Ca^{2+}, cAMP) via gap junctions.

Every cell in a multicellular organism is bombarded by hundreds of signaling molecules. The cell responds selectively according to its differentiation or specialization during development and carries out special functions. For survival, most cells are programmed to depend on a specific set of signals. If these are not available, the cell activates a suicide program called programmed cell death or apoptosis. It is by taking advantage of this principle that cancer cells can be ablated. Most signaling molecules generally act in combinations, and the response of a group of cells can be controlled by a limited diversity of signaling molecules. Each cell population loses a small segment of telomere after each passage, ultimately reaching a stage of senescence and termination (Lundblad and Wright 1996).

The concentration of a short-lived signaling molecule can rapidly be adjusted. In most adult tissues, when the signal level is low the response fades away. Many intracellular regulatory molecules have half-lives of less than 10 min. Covalent modification of proteins, e.g., phosphorylation and dephosphorylation of amino acids, tyrosine, serine, and threonine, occurs quickly. Nitric oxide (NO) made by the enzyme NO synthase by deamination of l-arginine has a short half-life of 5–10 s and is converted to nitrites and nitrates. NO reacts with the iron moiety of its receptor enzyme guanylyl cyclase to produce cAMP (Halliwell et al. 1996). The small hydrophobic molecules, e.g., steroid hormones (cortisol, estradiol, testosterone), thyroid hormones (thyroxine), retinoids, and vitamin D, diffuse through the cell membrane and act on a structurally related superfamily of intracellular receptors. Each receptor has a hormone-binding site at the carboxy terminus, a transcription-activating domain at the amino terminus, and a DNA-binding domain in the middle with a pocket at the N terminus and midsection to bind the inhibitor. The inhibitor dissociates from the receptor upon ligand binding and exposes the DNA-binding site. Each receptor binds DNA either as homodimers or as heterodimers. Unlike water-soluble molecules, the effect of steroid hormones lasts for hours in the blood. These hormones bind to DNA sequence adjacent to the gene that they regulate. Ligand binding alters the conformation of the receptor, which either activates or suppresses the gene within a short time (30 min). The primary response lasts for only a short time and the secondary response lasts for several days.

The three types of cell surface receptors are well characterized; they are either ion-channel-linked, G protein linked, or enzyme linked. Recombinant DNA techniques using gene cloning have permitted large-scale production and characterization of these proteins, which are present in trace amounts in cell membrane (0.01% of total protein mass). These diverse receptor proteins can be grouped together in a small number of superfamilies. The ion channel-linked receptors are involved in rapid synaptic signaling between electrically excitable cells. G protein-linked receptors indirectly regulate the activity of a separate plasma membrane target protein, either an enzyme or an ion-channel. Interaction between the receptor and target protein is mediated by a third protein, a trimeric GTP-binding regulatory protein (Fig. 2A,B). These G proteins belong to a large superfamily of seven-pass transmembrane proteins and are composed of three polypeptide chains (α, β, and γ). $G_S\alpha$ peptide binds and hydrolyzes GTP and activates adenyl cyclase, while $G_S\beta$ and $G_S\gamma$ peptides form a tight complex and anchor G_S peptide to the cytoplasmic side of membrane by a prenyl group. In the resting state of the G protein, $G_S\alpha$ peptide binds GDP. Upon activation by a ligand-bound receptor, $G_S\alpha$ peptide exchanges GDP for GTP and $G_S\alpha$ peptide dissociates from the βγ complex. In this case, the protein gains an extra phosphate group indirectly via GTP binding.

Upon activation, enzyme-linked receptors function directly as an enzyme or are associated with other enzymes. Most of them are single-pass transmembrane proteins; the ligand-binding side lies outside and the catalytic site inside the cell membrane. The enzyme-linked receptors are fairly heterogenous and are either protein kinases or associated protein kinases that phosphorylate specific sets of downstream proteins, thus generating a phosphorylation cascade in the target cells. Two types of kinases are known: tyrosine kinase, which phosphorylates on tyrosine residues, and serine–threonine kinase, which phosphorylates predominantly on serine and occasionally on threonine residues. A few kinases can phosphorylate all three residues. A group of intracellular proteins have been found, including GTPase-activating protein (GAP), phospholipase C-γ (PLC-γ) in the inositol phospholipid signaling pathway, and Src-like nonreceptor protein tyrosine kinases, that bind to tyrosine-P on activated receptor kinases. These adaptor proteins usually share two highly noncatalytic domains, called Src homology regions 2 and 3 (SH2 and SH3). They were first found in Src protein of cancer cells. One such SH-adaptor protein was found during genetic studies on the nematode worm (*sem*-5 in *C. elegans*). Mutations in the *sem*-5 gene block the signaling pathways from several receptor kinases and significantly affect the development of this worm.

12.10 Protein Kinase Superfamily

The protein kinase family includes more than a thousand diverse, but related enzymes that regulate all aspects of cellular growth, development, differentiation, and cell division. The first phosphorylase kinase was found in 1956 by Krebs and Fisher, who were awarded the Nobel prize for this accomplishment. This was followed by cAMP-dependent protein kinase (PKA). The inactive holoenzyme is a heterotetramer containing both regulatory (R) and catalytic subunits (C). After PKA activation, the C subunit phosphorylates cytoplasmic substrates, migrates to the nucleus, and phosphorylates CREB protein, which is responsible for gene activation. The R subunit inhibits catalytic activity of the C subunit and prevents its nuclear entry by keeping it in the bound form. The crystal structure of the catalytic subunit of PKA complexed with a peptide inhibitor was determined by Knighton et al. (1993) at a resolution of 2.0 α. The PKA protein was found to be expressed in *E. Coli* by Slice and Taylor (1989). The general rules of its substrate binding and recycling also apply to other members of the protein kinase family (Parker 1994; Gould 1994).

Three isoforms of PKA and the corresponding genes are expressed in mammalian tissues. The most abundant Cα subunit is also expressed in almost all tissues; the Cβ subunit is expressed in neuronal tissues, and the Cγ subunit is found in the testes. These subunits are highly conserved in all vertebrates and are also found in nematodes. The conserved catalytic core extends from residues 40–300. It includes the ATP-binding domain (residues 40–127) encoded by exons III–V and a peptide-binding catalytic phosphorylating domain encoded by exons VI–IX. There is a strong preference for basic residues preceding the phosphorylation site. The site of phosphorylation of amino acid residue (phosphotransfer) is designated as the P site, with the P-2, P-3, and P-6 sites having a preference for arginine. PKA binds protein kinase inhibitor (PKI) with a high affinity (<n*M* range). The peptide specificity of the catalytic subunit of PKA at several recognition sites was deduced from the phosphorylation sites and by using peptide analogues. The alignment of myosin light chain kinase and $pp60^{src}$ with the elements of the catalytic subunits of PKA is based on the modeling and use of coordinates of the Cα subunit. PKC, cdc2, and epithelial growth factor (EGF) receptor is also modeled in this way. cdc2 alignment is based on its recently determined crystal structure (De Bondt et al. 1993). Eight of the nine strands in PKA are present in cdc2, and all of the α-helices in the core, with the exception of helix B, are also present in cdc2; like the β-strands, the helix bundle (D,E,F,H) is superimposable.

12.10.1 Inhibitors of Protein Kinase A

Like all the activation signals, most of the protein kinases are maintained in an inactive state by binding with other proteins. Two groups of inhibitor proteins are: the R subunits of PKA and heat-stable smaller PKI. R subunits bind PKA reversibly in the absence of cAMP, and PKIs inhibit PKA by tightly binding the catalytic subunit. Four unique genes are found to code for R subunits and can be classified in two groups: the R_I subunits ($R_I\alpha$ and $R_I\beta$) are not autophosphorylated, whereas the R_{II}

subunits ($R_{II}\alpha$ and $R_{II}\beta$) are. All of the R subunits have a well-defined domain structure (Fig. 1) with variable and conserved segments. The carboxy terminus includes two tandem cAMP-binding domains. Each R subunit and that of PKI has an inhibitor site; mutation in this region interferes with holoenzyme formation. Several isoforms of PKI (PKIα and PKIβ) have been identified and sequenced; like the R subunits of PKA, they have high-affinity Mg-ATP-binding sites (Berridge 1991).

12.10.2 Linkage of Receptor Protein Kinases to Ras Protein Superfamily and Downstream Mitogen-Associated Protein Kinases

The Ras, Rho, Rac, and Rab subfamilies belong to the large Ras superfamily of monomeric GTPases and contain a prenyl group for membrane anchoring. These proteins function as switches and relay signaling from the membrane to the nucleus in order to stimulate transcription, proliferation, or differentiation. In 30%–35% of human cancers, mutations of *ras* gene were found. In the presence of dominant mutant *ras* or after microinjection of neutralizing anti-*Ras* antibodies, the stimulation response of receptor tyrosine kinase is blunted, and proliferation does not occur in mutant or treated cells. Ras hydrolyzes GTP 100 times more slowly than $G_s\alpha$ peptide; however, in the GTP-bound state, Ras is constantly active and sends proliferation signals. Two classes of signaling proteins affect Ras function. GAPs increase hydrolysis of *ras*-bound GTP and hence inactivate Ras, and guanine nucleotide-releasing proteins (GNRPs) counteract GAP activity by promoting the exchange of GTP for GDP. Ras and *ras*-regulating proteins have been highly conserved in evolution from the fruitfly to the worm and to humans.

12.10.3 Activation of Mitogen-Associated Protein Kinase Cascades by Ras

Ras activtion by phosphorylation is short-lived and quickly reversed by dephosphorylation. However, for proliferation and differentiation, the short-lived signals are converted into longer-lasting events that can sustain the signal and relay it to a specific gene in the nucleus via the transcription factor. MAP kinase and many serine/threonine kinases perform this important function; since they are also switched on by extracellular proliferation and differentiation signals and propagate signals via receptor tyrosine kinases and G protein-linked receptors, they are also known as extracellular signal-regulated kinases (ERKs). Full activation requires both tyrosine-P and threonine-P. The protein kinase that catalyses the P additions is called MAP kinase kinase. MAP kinase kinase itself is catalyzed by MAP kinase kinase kinase after Ras binding. Transcription of immediate-early genes is activated within minutes after growth factor binding to membrane growth hormone-releasing factor (GRF) receptor. Serum response factor (SRF) is constitutively bound to specific DNA sequence (SRE) of *fos* and a few other early response genes. Activated MAP kinase migrates from the cytosol to the nucleus and phosphorylates Elk-1 to activate the *fos* gene. MAP kinase phosphorylates Jun protein, which in turn combines with Fos protein to form the active GRP, AP-1, which activates other genes.

12.11 Cell Signaling in Cell Cycle and DNA Replication

Mammalian cells have a similar cell cycle control system with a major checkpoint at gap 1 (G_1) and remain quiescent in a state called G_0, until they receive growth factor signals from other cells.

Most cells have to be anchored to the extracellular matrix before cell division. The proteins in the signaling cascades were identified in studies of cancer cells. Mutations of proto-oncogenes, e.g., *Myc*, produce proteins that activate genes for proliferation.In contrast, the products of tumor suppressor genes, e.g., Rb, block proliferation by binding the GRPs. In addition to short-term regulation of proliferation, long-term regulatory mechanisms also exist, which affect the loss of proliferation potential in senescent cells or other regulatory mechanisms of higher growth and cell division during embryonic development, growth, and differentiation (Murray and Hunt 1993).

12.12 Restenosis, Endothelial Damage, and Early Gene Response

A number of growth factor genes are induced to direct and promote tissue repair. Khachigian et al. (1996) showed that, after injury to rat aorta, there is a marked increase of early growth response gene product (Egr-1) at the wound edge. Egr-1 interacts with the proximal promoter for the platelet-derived growth factor B chain (PDGF-B) gene and displaces the transcription factor Sp1 from the same binding site. Dzau and others (Dewanjee 1997; Finkel and Epstein 1995; Morishita et al. 1994; Nabel and Nabvel 1996; Pickering et al. 1996; Robinson et al. 1996) found that several ASONs inhibit transcription of proteins involved in regulating the cell cycle; they demonstrated that soaking saphenous vein with ASON specific for E_2F, cyclins, and cdks reduced intimal proliferation after vein grafting.

12.13 Dependence of Tyrosine Kinase-Associated Receptors on Nonreceptor Tyrosine Kinases

Many cell surface receptors lack catalytic domains and are not G protein linked; these receptors belong to the Src and Janus family of nonreceptor tyrosine kinases. They function like the receptor tyrosine kinases, but are noncovalently associated with a kinase domain encoded by a different gene. The cytokine receptors that regulate proliferation and differentiation in the hemopoietic systems and antigen-specific receptors on T and B lymphocytes. Several members of the Src family (e.g., Src, Yes, Fgr, Fyn, Lyn, Hck, Blk) contain SH2 and SH3 domains on the cytoplasmic sides. Receptor tyrosine kinases and Src family tyrosine-associated receptors activate the same signaling pathways. The Janus family of nonreceptor kinases (e.g., JAK1, JAK2, JAK3, JAK4) play an important role in inflammation and infection (Briscoe et al. 1996). Ligand binding in many tyrosine kinase-associated receptors causes the assembly of transmembrane receptor subunits, e.g., IL-2 secreted by T lymphocytes stimulates lymphocyte proliferation; the IL-2 receptor subunits (α, β, and γ) assemble on membrane after IL-2 binding. Neuronal cells divide only after injury or when they are activated by a growth factor. In contrast, cancer cells proliferate excessively due to the accumulation of several mutations, whereby they are able to multiply without stimulatory signals and are able to avoid the checkpoints of normal cell control. Many of these gene mutations affect encoded proteins in the signaling pathways.

Mutant genes were termed *oncogenes* before their origin from normal genes was understood; normal genes are known as *proto-oncogenes*. Viruses, radiation (ultraviolet, X-, and γ-rays), and chemical carcinogens induce germline and somatic mutations and activate normal proto-oncogenes to several transforming oncogenes. As already mentioned, in vivo carcinogenesis is a multistep process; two or more activated oncogenes are necessary to induce cell transformation and proliferation. Among the 70 known oncogenes (mutated genes), about 12 are also found in human cancer. They are revealed by their presence in transforming viruses or by their ability to transform 3T3 cells (murine fibroblast, NIH) in transfection assays. The cellular proto-oncogenes are divided into seven families:

- Group I consists of protein kinases (e.g., src, yes, abl, fps, erbB, fms, ros, kit, mos), which mainly phosphorylate tyrosine residues and a few of which phosphorylate threonine or serine residues
- Group II comprises the GTP-binding proteins (H-ras, K-ras)
- Group III contains growth factor (*sis* gene coding for PDGF)
- Group IV contains polypeptide growth factor receptor (erbA)
- Group V consists of nuclear proteins (myc, myb, fos, ski)
- Group VI comprises cyclin gene (D_1) products, which drive the cell cycle at a faster pace
- Group VII is made up of those proto-oncogenes that do not belong to one of the other six groups (rel, ets)

Most of these oncogenes work synergistically to promote cell proliferation and are activated via genetic alterations in a variety of human tumors (Stehelin et al. 1976; Watt et al. 1983; Lange-Carter et al. 1993; Janssen-Durr et al. 1993; Ohtsubo and Roberts 1993; Hesketh 1994). Estrogen treatment activates c-*myc*, c-*fos*, and c-*jun* oncogenes in rat uterus (Weinberg 1996). The protein products of the c-*fos* and c-*jun* genes form a heterodimeric complex that interacts with the gene promoter consensus sequence termed AP-I. Similarly, MYC protein binds with MAX protein (MYN in the mouse) and modulates genes via a different consensus sequence. This MYC–MAX complex also inactivates the tumor suppressor gene Rb (Blackwood and Eisenman 1991). ASON complementary to c-*myc* mRNA inhibits estrogen induction of breast cancer (Kinzler and Vogelstein 1996).

The cell continuously integrates input signals and generates a response for living, dying, proliferation, or repair induced by damage or extracellular stress. As more quantitative data about signaling become available, computer simulations will play an important role in elucidating the mechanism of decision-making in the signaling pathways. All the intracellular signaling pathways together form a highly interconnected network in which signals are processed along multiple parallel routes interacting with one another. Computer-based networks, called neural networks, have been developed to understand the relay of the neuronal message and information processing (Alberts et al. 1994). The simple type of neural networks consists of three layers of interconnected units: an input layer, a hidden layer, and an output layer. Each unit in the network functions as a unit neuron. The connections between units are similar to those of synapses, and modifiable weighting factors exist that control the strength with which one unit influences another. The activity of the intact network depends on the weighting factors and on mathematical rules (Markov principles), transforming input patterns into a different pattern of output units via the hidden connections. These networks can be trained to recognize specific input patterns; this is achieved by presenting the networks with training examples in the form of a series of inputs together with a desired pattern of outputs. The objective is to modify the weighting factors so that their performance will improve on subsequent presentation of the training set to yield the desired output. Once the network has learned the desired task, it can recognize and give the correct output for novel input patterns. The hypothetical network may consist of several receptors for growth factors and matrix molecules that will activate several MAP kinase kinases to converge on a single MAP kinase to activate transcription.

For a unicellular organism, about 100–200 genes are sufficient for cell survival and routine homeostatic functions. However, for the survival and functions of development and differentiation of multicellular mammalian organisms, such as humans, 1000 times more genes and redundancy are required for communication that is more effective and faster than invading and competing organisms or other pathogens. To protect the genes from multiple oxidative inactivating reactions, DNA exons are protected by introns and a large number of other trivial genes. The redundant systems also slow down the desired reaction rate by undesirable encounters. Most of the important cellular functions can be explained in terms of chemical reactions, e.g., addition of a phosphate or its removal after the signaling is over and the message has been relayed. This suggests that, for every exquisite function at any given time in the signaling cascade, enough energy must be spent to increase the concentration of reacting chemicals above the threshold value; this concentration requirement may be even higher to push the forward reactions for the creative forces before entropy dilutes and diffuses these reactants. Evolution, survival, and creativity are very selfish, single-minded, and capitalistic paradigms. It is important to develop techniques to identify radiolabeled substrates of transfected genes, e.g., thymidine kinase, to follow up the fate of transfected genes. However, it is more also important to develop probes that permit the activation of inducible and constitutive genes to be identified noninvasively.

So far, 6000–8000 mammalian species have been identified. In an adult human, there are about 40 trillion cells, including 100 billion neurons. In the human genome, there are about 80000–100000 genes, with similar numbers of mRNAs and an even greater number of proteins. Currently, about 1000 superfamilies of proteins have been identified for 100000 or more of these proteins; on the average this amounts to ten proteins or genes for each family of protein. In reality, redundancies exist for some essential proteins and not for some other proteins. The redundancy of genes for the same function increases almost fourfold, from fruit fly to *Homo sapiens* as a safety margin. The gene knockout mice involving 1000 genes suggest that for essential genes, the fetus does not survive till birth. In this redundant environment, the inhibition of protein expression by ASONs may be lower than expected (Alberts et al. 1994).

Gene sequencers spin out 5–6 million nucleotides a day; it is critical to recognize the functional and disease-causing genes in these nucleotide sequences. The key to determining the function of proteins encoded by these genes is to find related genes and proteins whose functions are known in bacteria, yeast, nematodes, and mammalian cells. Throughout evolution, a small number of sequences have been used over and over again in different settings. Using fast computers, genetists should be able to search genomes of different species more rapidly in order to derive information about genomes and protein functions. Programs such as BLAST and FASTA can find matches for 40%–50% of all new protein and gene sequences. In order to move into the twilight zones of sequence similarities, neural networks using the hidden Markov model (HMM) have been found to be very useful, increasing the chance of a homology search by another 10%–15%. The biology workbench developed at the National Center for Supercomputing Applications (NCSA) at the University of Illinois permits an unknown nucleotide sequence to be aligned, using applet software, by matching it against sequences in the GenBank database at NIH (National Center for Biotechnology Information, NCBI) and Heidelberg, the protein database at Brookhaven and Geneva, and the combined database ENQuire (30 databases and 100 analysis tools), using the Java language station and Web browsing. This and other tools permit the analysis of nucleotide and amino acid sequences along with the structure and functions of genes and proteins. The interface is written in Web's vernacular, hypertext markup language, or HTML (Fischman 1996; Milner et al. 1997)). The goal of genomics in drug discovery is to develop critical targets using bioinformatics. Combinatorial chemistry programs are generating a wide range of ligands and rely on the growing field of chemoinformatics to control and guide this diversity. Scaled-up screening processes merge critical targets with specific ligands and help to identify lead compounds in drugs and radiopharmaceuticals.

12.14 Interactions Between Antisense Oligonucleotides and mRNA Targets

12.14.1 Molecular Mechanism

The oligonucleotide effect can be rationalized by its interaction with the cytoplasmic mRNA target functioning as a receptor. The ASONs may inhibit gene expression by translational arrest, activation of RNaseH, and oligonucleotide cleavage of mRNA. However, in their rational design as pharmacological agents, several differences in affinity, specificity, and nonspecific actions have also been observed.

12.14.2 Parameters of Affinity in Oligonucleotide–mRNA Hybridization Reactions

The affinity of oligonucleotide for mRNA sequence depends on the hybridization of complementary sequences of neighboring bases. Two determinant factors for ASON–mRNA affinity are:

1. Free energy of hydrogen bonding from Watson–Crick base-pairing
2. Base stacking of the asymmetric mRNA–single-stranded DNA to form a double helix

In in vitro systems, the affinity of the negatively charged oligonucleotide to mRNA can be enhanced at higher ionic strength. Affinity also increases with the length of the oligonucleotide and is highly dependent on the nucleotide sequence. The free energy of binding of DNA–DNA, RNA–RNA, or DNA–RNA hybrid has been estimated (Freier et al. 1992).

12.14.3 Specificity of Hybridization Reactions

The specificity of hybridization reactions results from the interactions of adenine with thymidine and of guanine with cytidine. Loss of affinity from mismatched base pairs depends on the type of specific mismatch, the position of the mismatch in the region of complementarity, and the neighboring sequences. The change in the free energy of binding by a single mismatch is of the order of 0.2–4.0 kcal/mol at 100 m*M* ionic strength. A single base mismatch might drastically change affinity by a factor of 100–500. Statistical calculations suggest that a unique sequence for every gene can be obtained by including at least 17 nucleotides for the intact human genome (Thein and Wallace 1986). For the nonrandom RNA population, a unique sequence can be obtained by including at least 13 nucleotides (Helene and Toulme 1990). To optimize sequence selection, the length and sequence must be carefully manipulated by checking for tertiary structure of mRNA, accessibility of hybridization, and the possibility of self-folding. For hybridization with a mutant mRNA, the exact site of mutation and base must be identified, and a higher level of hybridization (four- to fivefold) with the mutant can thus be attained over the normal mRNA sequence. Although ionic strength, concentration, and other hybridization parameters can be adjusted with cultured cell lines in in vitro experiments, in vivo hybridization and theoretical specificity can only be optimized by several trial runs. The strategy for diagnosis will be different than for therapy. For *diagnosis*, we need amplified copies of the mRNA target to generate image contrast. For *therapy*, we need the target at the critical checkpoint, e.g., reduction of a specific protein initiating transcription factor (c-myc) or holding the cell cycle at arrest (E2F), which could block downstream signaling for transcription and DNA replication (Morishita et al. 1994).

12.15 Potential Application of Ribozyme in Inhibition of Transcription

Ribozymes, e.g., small, hammerhead-shaped RNA molecules, have been used as antisense oncogenes and may provide specific delivery for a short sequence of 40–50 bases (Scanlon et al. 1994; Lu et al. 1994; Rossi 1995; Roush 1997; Anonymous 1997). Their efficiency is two to three orders of magnitude lower than that of an enzyme. However, with improved design, they may find novel applications.

12.16 Applications of Phosphorothioate Oligonucleotides in Diagnosis and Therapy

Among the first-generation oligonucleotide derivatives, the phosphorothioate analogues have been extensively evaluated in several animal models and in several clinical trials and demonstrated superior therapeutic effects. However, several limitations were also observed, e.g., a lower affinity per nucleotide, thus requiring a longer nucleotide sequence and less unwinding of mRNA structures. At higher concentrations, they also were found to inhibit RNaseH (Helene and Toulme 1990; Wagner 1994; Frier et al. 1992; Bennett et al. 1994). The dose–response curve was found to be biphasic, especially in the high-dose range. In addition, they do not cross blood–brain barrier. Due to sensitivity of degradation in variable ranges of acidity and alkalinity of the alimentary tract, they are not orally bioavailable. They also showed dose-dependent pharmacological effects. An adverse effect was observed due to activation of monocytes and neutrophils, release of cytokines, and activation of clotting factors and complement.

12.17 Chemistry of Modified Oligonucleotides

Considering the sensitivity of phosphodiester linkage to nuclease attack, several modifications were made to make them nuclease resistant. These modifications were made by changing one of the following three sites:

1. Phosphodiester backbone
2. Sugar moiety
3. Purine or pyrimidine bases

By tinkering with the 2'-position of the deoxyribosyl moiety of sugar with a methoxyethoxy group, Cook and collaborators at ISIS and Ciba-Geigy Inc. found that these modifications provide nuclease-resistant, tightly binding oligonucleotides (Beaucage and Iyer 1993; Wagner et al. 1993; Kawasaki et al. 1993; Cummins et al. 1995; Arnold 1995). However, the mRNA–oligonucleotide duplex are not recognized by the RNaseH enzyme, and mRNA and protein production are not inhibited. Hence, another modification called "gapmer strategy" was used, in which the midportion of oligonucleotide has RNaseH recognizing the segment of oligonucleotide-mRNA hybrid (Schein 1997). The gapmers have 60% less sulfur and bind more tightly than phosphorothioate; they also have better antitumor activity, desirable tissue distribution, and fewer side effects, thus providing the benefits of oligonucleotide drugs at lower concentrations (Fig. 13A,B).

12.18 Adverse Effects of Administration of Oligonucleotides

Several oligonucleotides, particularly phosphorothioate derivatives, have been evaluated in animal models and human patients. The marked toxicity observed in rodent species was attributed to the release of cytokines (Breslauer et al. 1986). Activation of complement and clotting factors and hypotension was observed in primate models (Crooke and Bennett 1996). About a dozen phosphorothioate oligodeoxynucleotides are currently undergoing clinical trials and have been well tolerated in the dose ranges used (2 mg/kg intravenously over a duration of 2 h). ISIS 2105 is directly applied onto genital warts induced by human papillomavirus (Cowsert et al. 1993), and ISIS 2922 is administered intravitreously in AIDS patients with cytomegalovirus-induced retinitis (Lisziewicz et al. 1994). GEM91 is designed to target human immunodeficiency virus (HIV) Gag mRNA and is administered intravenously (Lisziewicz et al. 1992). ISIS 2302 is designed to inhibit intercellular adhesion molecule (ICAM-1) of endothelial cells in order to reduce inflammation (Bennett et al. 1994). An oligonucleotide targeting c-myb oncoprotein has been evaluated for leukemia treatment (Ratajczak et al. 1992). Some oligomers were found to activate neutrophils and monocytes and to induce pulmonary margination, complement activation, and activation of enzymes in the clotting cascade.

As information technology is developed for homology search of available genome databases, the barriers and limitations of searching and finding unique short oligonucleotide probes will be overcome in the near future. Most of the oligonucleotides are synthesized for making primers for polymerase chain reactions. For in vivo applications, more nuclease-resistant probes need to be developed by modifying the phosphodiester backbone, sugar, or base of nucleotides, while retaining the desirable properties of membrane permeability and hybridization. Oligonucleotide synthesis is highly automated, and postsynthesis modifications are easy to carry out. The final step of amino-linking and conjugation may be incorporated in the final step of automated synthesis. Abundance, stability, and availability of an mRNA target are issues that need to be addressed in the design of each probe. As the pharmaceutical industry becomes more successful in developing therapeutic oligonucleotide drugs and as instruments for large-scale production at a reduced cost become available, these probes will develop to maturity. The main focus is on oncogenes and viral genes. It is encouraging to note that Paul Zamecnik was awarded the Laskar award in 1996 for the original work he carried out on the antisense mechanism in 1976. We are optimistic that novel diagnostic and therapeutic applications using radiolabeled antisense oligomer mechanisms will follow in the near future.

Acknowledgements. The support of the National Heart, Lung and Blood Institute (HL-47201) and the American Cancer Society is gratefully acknowledged.

References

Agrawal S, Temsamani J, Tang JY (1991) Pharmacokinetics, biodistribution and stability of oligodeoxynucleotide phosphorothioate in mice. Proc Natl Acad Sci 88:7595–7599

Akhtar S, Agrawal S (1997) In vivo studies with antisense oligonucleotides. Trends Pharmacol Sci 18:12–18

Alberts B, Bray D, Lewis J, Raff M, Roberts K, Watson JD (1994a) Cell signaling. In: Alberts B, Bray D, Lewis J, Raff M, Roberts K, Watson JD (eds) Molecular biology of the cell. Garland, New York, pp 721–785

Alberts B, Bray D, Lewis J, Raff M, Roberts K, Watson JD (1994b) The cell division cycle. In: Alberts B, Bray D, Lewis J, Raff M, Roberts K, Watson JD (eds) Molecular biology of the cell. Garland, New York, pp 863–910

Anonymous (1997) Antisense '97: a roundtable on the state of the industry. Nat Biotechnol 15:529–536

Arnold LJ (1995) The developments of oligonucleotides containing chirally pure methylphosphonates as therapeutic agents. Antisense Res Dev 5:108

Avery OT, MacLeod CM, McCarty M (1944) Studies of the chemical nature of the substance inducing transformation of pneumococcal types. J Expt Med 79:137

Baum R (1994) Chemists create new oligonucleotide anlogs for antisense applications. Chem Eng News 21–22

Beaucage SL, Iyer RP (1993) The synthesis of modified oligonucleotides by the phosphoroamidite approach and their applications. Tetrahedron 49:6123–6194

Beltinger C, Saragovi HU, Smith RM, LeSauteur L, Shah N (1995) Binding, uptake and intracellular trafficking of phosphorothioate-modified oligonucleotides. J Clin Invest 95:1814–1823

Bennett CF, Crooke ST (1996) Regulation of endothelial cell adhesion molecule 1 (ICAM-1) expression with antisense oligonucloetides. Adv Pharmacol 28:1–43

Bennett CF, Condon T, Grimm S, Chan H, Chiang M-Y (1994) Inhibition of endothelial leukocyte adhesion molecule expression with antisense oligonucleotides. J Immunol 152:3530–3540

Benton WD, Davis RW (1977) Hybridization in situ of SV40 plaques: detection of recombinant SV40 virus carrying specific sequences of non-viral DNA. Science 196:180–182

Berridge MJ (1991) The molecular basis of communication within the cell. In: Rasmussen H (ed) Cell communication in health and disease. Scientific American, Freeman, New York, pp 37–50

Bjorklund S, Kim Y-J (1996) Mediators of transcriptional regulation. Trends Biochem Sci 21:335–341

Blackwood EM, Eisenman RN (1991) Max: a helix-loop-helix zipper protein that forms a sequence-specific DNA-binding complex with myc. Science 251:1211–1217

Blatteis CM (1997) Crystal structure of a peptide nucleic acid (PNA) duplex at 1.7 A resolution. Nat Struct Biol 4:98–100

Breslauer KJ, Frank R, Blocker H, Marky LA (1986) Predicting DNA duplex stability from base sequence. Proc Natl Acad Sci 83:3746–3750

Briscoe J, Kohlhuber F, Muller M (1996) JAKs and STATs branch out. Trends Cell Biol 6:336–340

Brown DA, Kang SH, Gryaznov SM, DeDinioso L, Heidenreich O (1994) Effect of phosphorothioate modification of oligonucleotides on specific protein binding. J Biol Chem 269:26801–26805

Buongiorno-Nardelli M, Amaldi F (1970) Autoradiographic detection of molecular hybrids between rRNA and DNA in tissue sections. Nature 225:946

Burgess TL, Fisher EF, Ross SL, Bready JV, Qian Y-X, (1995) The antiproliferative activity of c-myc and c-myb antisense oligonucleotides in smooth muscle cells is caused by a nonantisense mechanism. Proc Natl Acad Sci USA 92:4051–4055

Campuzano V (1996) Friedreich's ataxia: autosomal recessive disease caused by an intronic GAA triplet repeat expansion. Science 271:1423–1427

Chiasson BJ, Hooper ML, Murphy PR, Robertson HA (1992) Antisense oligonucleotide eliminates in vivo expression of c-fos in mammalian brain. Eur J Pharmacol 227:451–453

Chollet A, Kawashima EH (1985) Biotin-labeled synthetic oligonucleotides: chemical synthesis and uses as hybridization probes. Nucleic Acid Res 13:1529–1541

Cowsert LM, Fox MC, Zon G, Mirabelli CK (1993) In vitro evaluation of phosphorothioate oligonucleotides targeted to the E2 mRNA of papilloma virus: potential treatment of genital warts. Antimicrob Agents Chemother 37:171–177

Crooke ST, Bennett CF (1996) Progress in antisense therapeutics. Annu Rev Pharmacol Toxicol 36:107–129

Cummins LL, Owens SR, Risen LM, Lesnik EA, Freier SM (1995) Effects of uniform 2' modifications on oligonucleotide hybridization and nuclease sensitivity. Nucleic Acid Res 23:2019–2024

De Bondt HL, Rosenblatt J, Jancarik J, Jones HD, Morgan DO, Kim SH (1993) Crystal structure of cyclin-dependent kinase 2. Nature 363:595–602

Denu JM, Stuckey JA, Saper MA, Dixon JE (1996) Form and function in protein dephosphorylation. Cell 87:361–364

Dewanjee MK (1993) Radiololabeled antisense probes: diagnosis and therapy. Diagn Oncol 3:189–208

Dewanjee MK (1995) Quality assurance of radiolabeled proteins, peptides and antisense oligodeoxynucleotides. In: Emran A, Boothe T (eds) Chemist's view of imaging center. American Chemical Society Symposium. Plenum, New York, pp 189–210

Dewanjee MK (1996) Inter- and intracellular communications. In: Wagner HN, Szabo Z, Buchanan J (eds) Principles of nuclear medicine, 2nd edn. Saunders, Philadelphia, pp 28–39

Dewanjee MK (1997) Radiolabeled DNA probes: diagnosis and therapy. CRC Press, Boca Raton

Dewanjee MK, Serafini AN (1996) Molecular biology of engineered proteins. In: Torchilin V (ed) Designer antibodies as targeting diagnostic units. CRC Press, Boca Raton, pp 46–75

Dewanjee MK, Ghafouripour AK, Werner RK, Serafini AN, Sfakianakis GN (1991) Development of sensitive radioiodinated anti-sense oligonucleotide probes by conjugation technique. Bioconjugate Chem 2:195–200

Dewanjee MK, Ghafouripour AK, Kapadvanjwala M, Oates E, Krishan A, Lopez DM, Serafini AN, Sfakianakis GN (1993a) Radiolabeled antisense oligodeoxynucleotides: in vitro and in vivo applications. J Clin Immunoassay 16:276–289

Dewanjee MK, Ghafouripour AK, Subramanian M (1993b) Labeling antisense deoxyoligonucleotide with Tc-99m and hybridization with c-myc oncogene mRNA in P388 leukemic cells. 10th international symposium of radiopharmaceutical chemistry, Kyoto, pp 40–42

Dewanjee MK, Ghafouripour AK, Kapadvanjwala M, Dewanjee S, Serafini AN, Lopez DM, Sfakianakis GN (1994a) Noninvasive imaging of c-myc oncogene messenger RNA with In-111 labeled antisense probes in mammary tumor-bearing mouse model. J Nucl Med 35:1054–1063

Dewanjee MK, Ghafouripour AK, Kapadvanjwala M, Samy AT (1994b) Kinetics of Hybridization of mRNA of c-myc oncogene with In-111 labeled antisense oligodeoxynucleotide probes by high pressure liquid chromatography technique. Biotechnique 16:844–850

Dewanjee MK, Ghafouripour AK, Mallin W, Serafini AN, Sfakianakis GN, Werner RK, Samy AT, Krishan A, Glenn SD, Gupta RK, Subramanian M, Hanna MG (1994c) Direct method of cell-labeling in whole blood by administration of In-111 labeled antisense oligonucleotide probe: evaluation of radiolabeled probes for noninvasive imaging of inflammation and infection. In: Martin-Comin J (ed) Radiolabeled blood elements congress of International Society of Radiolabeled Blood Elements, Barcelona, Spain, 12–16 Nov 1992. Plenum, New York, pp 7–13

Dewanjee MK, Abrams MJ, Gonzalez L, Dewanjee S, Serafini A, Sfakianakis GN (1994) Radiolabeled single-stranded DNA probe for imaging thrombin in thrombus. Noninvasive imaging of DNA-protein interaction. J Labeled Compounds Radiopharm 37:347–349

Dewanjee MK, Ghafouripour AK, Hanna M, Subramanian M, Kapadvanjwala M, Lopez DM, Serafini AN, Sfakianakis GN (1995) Tumor cell killing by Y-90 labeled Bcl2 antisense oligodeoxynucleotide probes. J Labeled Compounds Radiopharm 37:344–346

Duff JL, Monia BP, Berk BC (1995) Mitogen-activated protein (MAP) kinase is regulated by MAP kinase phosphatase (MKP-1) in vascular smooth muscle cells. Effect of actinomycin D and antisense oligonucleotides. J Biol Chem 270:7161–7166

Duke RC, Ojcius DM, Young JD-E (1996) Cell suicide in health and disease. Sci Am 80–87

Egholm M, Buchardt O, Christensen L, Behrens C, Freier SM (1993) PNA hybridizes to complementary oligonucleotides obeying the Watson-Crick hydrogen-bonding rules. Nature 365:566–568

Ellis HM, Horvitz HR (1986) Genetic control of programmed cell death in the nematode C. elegans. Cell 44:817–822

Ellis RE, Yuan JV, Horvitz HR (1991) Mechanisms and functions of cell death. Annu Rev Cell Biol 7:663–698

Fenster SD, Wagner RW, Froehler BC, Chin DJ (1994) Inhibition of human immunodeficiency virus type-1 envelope expression by C-5 propyne oligonucleotides specific for rev-response element stem-loop V. Biochemistry 33:8391–8398

Finkel T, Epstein SE (1995) Gene therapy for vascular disease. FASEB J 9(10):843–851

Fischman J (1996) Working the Web with a virtual lab and some Java. Science 273:591–593

Freier SM, Lima WF, Sanghvi YS, Vickers T, Zounes M (1992) Gene regulation: biology of antisense RNA and DNA. In: Erickson RP, Izant JG (eds) Thermodynamics of antisense oligonucleotide hybridization, vol 1. Raven, New York, pp 95–107

Gajewski TF, Thomson CB (1996) Apoptosis meets signal transduction: elimination of a BAD influence. Cell 87:589–592

Galbraith WM, Hobson WC, Giclas PC, Schecter PJ, Agarwal S (1994) Complement activation and hemodynamic changes following intravenous administration of phosphorothioate oligonucleotides in the monkey. Antisense Res Dev 4:201–206

Gall J, Pardue M (1969) Formation and detection of RNA-DNA hybrid molecules in cytological preparations. Proc Natl Acad Sci USA 63:378

Gao W-Y, Han F-S, Storm C, Egan W, Cheng Y-C (1992) Phosphorothioate oligonucleotides are inhibitors of DNA polymerases and RNase H: implications for antisense technology. Mol Pharmacol 41:223–229

Gewirtz AM, Stein CA, Glazer PM (1996) Facilitating oligonucleotide delivery: helping antisense deliver on its promise. Proc Natl Acad Sci USA 93(8):3161–3163

Gould KL (1994) Cyclin-dependent protein kinases. In: Woodgett JR (ed) Protein kinases. IRL Press, Oxford, pp 149–176

Grunstein M, Hogness DS (1975) Colony hybridization: a method for the isolation of cloned DNAs that contain a specific gene. Proc Natl Acad Sci USA 72:3961–3965

Hagiwara M, Brindle P, Harootunian AT, Armstrong R, Rivier J, Vale W (1993) The coupling of hormonal stimulation and transcription via cyclic AMP response factor CREB is rate limited by nuclear entry of protein kinase A. Mol Cell Biol 13:4852–4859

Halliwell B, Cross CE, van der Vliet A, O'Neill CA, Louie S (1996) The generation of oxygen-derived species. In: Weir EK, Archer SL, Reeves JT (eds) Nitric oxide and radicals in the pulmonary vasculature. Futura, Armonk, New York, pp 21–38

Helene C, Toulme J-J (1990) Specific regulation of gene expression by antisense, sense and antigene nucleic acids. Biochim Biophys Acta 1049:99–125

Hermanson GT (1996) Nucleic acid and oligonucleotide modification and conjugation. Bioconjugate techniques. Academic, New York, pp 640–671

Hesketh R (1994) Normal cell growth and functions of oncoproteins. In: Hesketh R (ed) The oncogene handbook. Academic, New York, pp 8–22, pp 312–352

Hijiya N, Zhang J, Ratajczak MZ, Kant JA, DeRiel K (1994) Biologic and therapeutic significance of MYB expression in human melanoma. Proc Natl Acad Sci USA 91:4499–4503

Holt J (1996) Antisense therapeutics. Mol Med Today 2 (5):184-185

Hooper ML, Chiasson BJ, Robertson HA (1994) Infusion in the brain of antisense oligonucleotide to the immediate early gene c-fos suppresses production of Fos and produces a behavioral effect. Neurosciene 63:917–924

Janssen-Durr P, Meichle A, Steiner P, Paganao M, Finke K, Botz J, Wessbecher J, Draetta G, Eilers M (1993) Differential modulation of cyclin gene expression by MYC. Proc Natl Acad Sci USA 90:3685–3689

John HA, Birnstiel ML, Jones KW (1969) RNA-DNA hybrids at the cytological level. Nature 223:582

Kaiser K, Meisterernst M (1996) The human general co-factors. Trends Biochem Sci 21:343–347

Kawasaki AM, Casper MD, Freier SM, Lesnik EA, Zounes MC (1993) Uniformly modified 2'-deoxy-2'-fluoro phosphorothioate oligonucleotide as nuclease-resistant antisense compounds with high affinity and specificity for RNA targets. J Med Chem 36:831–841

Kedes DH, Kedes LH (1996) Molecular genetics and its application to understanding cardiovascular disease. Dekker, New York, pp 1–78

Kerr JFR, Wyllie AH, Currie AR (1972) Apoptosis: a basic biological phenomenon with wide-ranging implications in tissue kinetics. Br J Cancer 26:239

Khachigian LM, Lindner V, Williams AJ, Collins T (1996) Egr-1-induced endothelial gene expression: a common theme in vascular injury. Science 271:1427–1431

Kindy MS (1994) NMDA receptor inhibition using antisense oligonucleotides prevents neuronal death in gerbil hippocampus following cerebral ischemia. Nature 363:260–263

Kinzler KW, Vogelstein B (1996) Lessons from hereditary colorectal cancer. Cell 87:159–170

Knighton DR, Bell SM, Zheng J, Ten Eyck LF, Xuong N-H, Taylor SS, Sowadski JM (1993) 2.0 A refined crystal structure of catalytic subunit of cyclic-AMP dependent protein kinase complexed with a peptide inhibitor and detergent. Acta Cryst D49:357–365

Konradi C, Cole RL, Heckers S, Hyman SE (1994) Amphetamine regulates gene expression in rat striatum via transcription factor CREB. J Neurosci 14:5623–5634

Korsmeyer SJ (1995) Regulators of cell death. Trends Genet 11:101–105

Krebs EG, Fisher EH (1956) The phosphorylase b to a converting enzyme of rabbit skeletal muscle. Biochim Biophys Acta 20:150–157

Krieg AM, Yi A-K, Matson S, Waldschmidt TJ, Bischop GA (1995) CpG motifs in bacterial DNA trigger direct B cell activation. Nature 374:546–549

Lange-Carter CA, Pleiman CM, Gardner AM, Blumer KJ, Johnson GL (1993) A divergence in the MAP kinase regulatory network defined by MEK kinase and Raf. Science 260:315–319

Larsen SK, Caldwell G, Higgins JD, Abrams MJ, Solomon HF (1994) Technetium complex of tricine: useful precursor for the ^{99m}Tc labeling of hydrazino nicotinamide modified proteins. J Labelled Compounds Radiopharm XXXV:1–2

Lisziewicz J, Sun D, Klotman M, Agrawal S, Zamecnik P, Gallo R (1992) Specific inhibition of immunodeficiency virus type 1 replication by antisense oligonucleotides: an in vitro model for treatment. Proc Natl Acad Sci USA 89:11209–11213

Lisziewicz J, Sun D, Weichold FF, Thierry AR, Lusso P (1994) An antisense oligonucleotide phosphorothioate complementary to Gag mRNA blocks

replication of human immunodeficiency virus type 1 in human peripheral cells. Proc Natl Acad Sci USA 91:7942–7946
Livesey FJ, Hunt SP (1996) Identifying changes in gene expression in the nervous system: mRNA differential display. TINS 19:84–88
Lodish H, Baltimore D, Berk A, Zipursky SL, Matsudaira P, Darnell J (1995) Cell-to-cell signaling: hormones and receptors. Scientific American, Freeman, New York, pp 853–924
Loke SL, Stein CA, Zhang XH, Mori K, Nakanishi M, Subasinghe C, Cohen JS, Neckers LM (1989) Characterization of oligonucleotide transport into living cells. Proc Natl Acad Sci USA 86:3473–3478
Lu D, Chatterjee S, Brar D, Wong KK Jr (1994) Ribozyme-mediated in vitro cleavage of transcripts arising from the major transforming genes of human papilloma virus type. Cancer Gene Ther 1 (4):267–277
Lundblad V, Wright WE (1996) Telomeres and telomerase: a simple picture becomes complex. Cell 87:369–375
Maltese JY, Sharma HW, Vassilev L, Narayanan R (1995) Sequence context of antisense RelA/NF-kB phosphorothioate determines specificity. Nucleic Acid Res 23:1146–1151
Manoharan M, Tivel KL, Andrade LK, Mohan V, Condon TP (1995) Oligonucleotide conjugates: alteration of the pharmacologic properties of antisense agents. Nucleoside Nucleotide 14:969–973
Martiny-Baron G, Marme D (1995) VEGF-mediated tumour angiogenesis: a new target for cancer therapy. Curr Opin Biotechnol 6 (6):675–680
Maxam AM, Gilbert W (1977) A new method of sequencing DNA. Proc Natl Acad Sci USA 74:560–564
McCarthy MM, Masters DB, Rimvall K, Schwartz-Giblin S, Pfaff DW (1994) Intracerebral administration of antisense oligonucleotides to GAD65 and GAD67 mRNAs modulate reproductive behaviour in the female rat. Brain Res 636:209–220
MilnerN, Mir KU, Southern EM (1997) Selecting effective antisense reagents on combinatorial arrays. Nat Biotechnol 15:537–541
Monia BP, Johnston JF, Ecker DJ, Zounes M, Lima WF, Frier SM. (1992) Selective inhibition of mutant Ha-ras mRNA expression by antisense oligonucleotides. J Biol Chem 267:19954–19962
Morishita P, Gibbons GH, Kaneda Y, Ogihara T, Dzau VJ (1994) Pharmacokinetics of antisense oligodeoxynucleotides (cyclin B_1 and CDC2 kinase) in the vessel wall in vivo: enhanced therapeutic utility for restenosis by HVJ-liposome delivery. Gene 149:13–19
Murray A, Hunt T (1993) The cell cycle: an introduction. Oxford University Press, Oxford
Nabel EG, Nabel GJ (1996) Gene-transfer approaches to vascular disease. Dekker, New York, pp 487–512

Nagata S, Golstein P (1995) The Fas death factor. Science 267:1449–1456
Narang SA, Jacobs TM, Khorana HG (1967) Studies on polynucleotides. LXIII. Deoxyribopolynucleotides containing repeating trinucleotide sequences. The polymerization of protected deoxyribonucleotides. J Am Chem Soc 89:2158–2177
Narayanan R, Higgins KA, Perez JR, Coleman TA, Rosen CA (1993) Evidence of differential functions of p50 and p65 subunits of NF-κB with a cell adhesion model. Mol Cell Biol 13:3802–3810
Nesterova M, Cho-Chung YS (1995) A single injection protein kinase A-directed antisense treatment to inhibit tumor growth. Nat Med 1:528–533
Ohtsubo M, Roberts JM (1993) Cyclin-dependent regulation of G_1 in mammalian fibroblasts. Science 259:1908–1912
Osen-Sand A, Catsicas M, Staple JK,Jones KA, Ayala G (1993) Inhibition of axonal growth by SNAP-25 antisense oligonucleotide in vitro and in vivo.Nature 364:445–448
Pandha HS, Lemoine NR (1996) Gene therapy in gastroenterology. Gut 38 (2):161–165
Parker PJ (1994) Protein kinase C and its relatives. In: Woodgett JR (ed) Protein kinases. IRL Press, Oxford, pp 68–84
Patikoglou G, Burley SK (1997) Eukaryotopic transcription factor-DNA complexes. Annu Rev Biophys Biomol Struct 26:289–325
Perez JR, Li Y, Stein CA, Majumder S, Van Oorschot A, Narayanan R (1994) Sequence-independent induction of Sp1 transcription factor activity by phosphorothioate oligonucleotides. Proc Natl Acad Sci 91:5957–5961
Pickering JG, Isner JM, Ford CM, Weir L, Lazarovits A, Rocnik EF, Chow LH (1996) Processing of chimeric antisense oligonucleotides by human vascular smooth muscle cells and human atherosclerotic plaque. Implications for antisense therapy of restenosis after angioplasty. Circulation 93 (4):772–780
Ratajczak MZ, Kant JA, Luger SM, Hijiya N, Zhang J (1992) In vivo treatment human leukemia in a scid mouse model with c-myb antisense oligonucleotides. Proc Natl Acad Sci USA 89:11823–11827
Robinson GS, Pierce EA, Rook SL, Foley E, Webb R, Smith LE (1996) Oligodeoxynucleotides inhibit retinal neovascularization in a murine model of proliferative retinopathy. Proc Natl Acad Sci USA 93 (10):4851–4856
Roeder RG (1996) The role of general initiation factors in transcription by RNA polymerase II. Trends Biochem Sci 21:327–334
Rossi JJ (1995) Therapeutic antisense and ribozymes. Br Med Bull 51 (1):217–225
Rous P (1911) Transmission of a malignant new growth by means of a cell-free filtrate. JAMA 56:198
Roush R (1997) Antisense aims for a renaissance. Science 276:1192–1193

Sagi J, Szemzo A, Ebinger K, Szabolcs A, Sagi G (1993) Base-modified oligonucleotides. I. Effect of 5-alkyl, 5-(1-alkenyl) and 5-(1-alkynyl) substitution of the pyrimidines on duplex stability and hydrophobicity. Tetrahedron Lett 34:2191–2194

Saleh M, Stacker SA, Wilks AF (1996) Inhibition of growth of C6 glioma cells in vivo by expression of antisense vascular endothelial growth factor sequence. Cancer Res 56 (2):393–401

Sambrook J, Fritsch EF, Maniatis T (1989) Synthetic oligonucleotide probes, chap 11. In: Sambrook J, Fritsch EF, Maniatis T (eds) Molecular cloning, a laboratory manual, vol 2. Cold Spring Harbor Laboratory Press, New York, pp 11.3–11.58

Sanger F (1981) Determination of nucleotide sequence of DNA. Science 214:1205–1210

Scanlon KJ, Ishida H, Kashani-Sabet M (1994) Ribozyme-mediated reversal of multidrug resistant phenotype. Proc Natl Acad Sci USA 91:11123–11127

Schein CH (1997) From housekeeper to microsurgeon: the diagnostic and therpeutic potential of ribonucleases. Nat Biotechnol 15:529–536

Schrump DS, Chen A, Consoli U (1996) Inhibition of lung cancer proliferation by antisense cyclin D. Cancer Gene Ther 3 (2):131–135

Shillitoe EJ, Kamath P, Chen Z (1994) Papillomaviruses as targets for cancer gene therapy. Cancer Gene Ther 1 (3):193–204

Simons M, Edelman ER, Rosenberg RD (1994) Antisense proliferating cell nuclear antigen oligonucleotides inhibit intimal hyperplasia in a rat carotid artery injury model. J Clin Invest 93:2351–2356

Slice LW, Taylor SS (1989) Expression of the catalytic subunit of cAMP-dependent protein-kinase in E. Coli. J Biol Chem 264:20940–20948

Sousa R (1996) Structural and mechanistic relationships between nucleic acid polymerases. Trends Biochem Sci 21:186–190

Southern EM (1975) Detection of specific sequences among DNA fragments separated by electrophoresis. J Mol Biol 98:503–517

Spolar RS, Record TM (1994) Coupling of local folding to site-specific binding of proteins to DNA. Science 263:777–784

Stamler JS, Dzau VJ, Loscalzo J (1996) The vascular smooth muscle cell. In: Loscalzo J, Creager MA, Dzau VJ (eds) Vascular medicine: a textbook of vascular biology and diseases. Little and Brown, Boston, pp 69–116

Stehelin D, Varmus HE, Bishop JM, Vogt PK (1976) DNA related to the transforming genes of avian sarcoma viruses is present in normal avian DNA. Nature 260:170

Stein CA, Cheng YC (1993) Antisense oligonucleotides as therapeutic agents – Is the bullet really magical? Science 261:1004–1012

Stepkowski SM, Tu Y, Condon TP, Bennett CF (1995) Induction of transplantation tolerance by treatment with ICAM-1 antisense oligonucleotides and anti-LFA-1 monoclonal antibodies. Transplant Proc 27:113

Sun J-S, Helene C (1993) Oligonucleotide-directed triple-helix formation. Curr Opin Struct Biol 3:345–356

Taylor SS, Radzio-Andzelm E (1994) Cyclic AMP-dependent protein kinase. In: Woodgett JR (ed) Protein kinases. IRL Press, Oxford, pp 1–29

Thein SL, Wallace RB. (1986) The use of synthetic oligonucleotides as specific hybridization probes in the diagnosis of genetic disorders. In: Davies KE (ed) Human genetic diseases: a practical approach. Oxford, IRL Press, pp 33–50

Tomei LD, Cope FO (1991) Apoptosis: the molecular basis of cell death. Cold Spring Harbor Laboratory Press, New York

Trauth BC, Klas C, Peters AMJ, Matzku S. Moller P, Falk W, Debatin _KM, Krammer PH (1989) Monoclonal antibody-mediated tumor regression by induction of apoptosis. Science 245:301–306

Viskers T, Baker BF, Cook PD, ZOunes M, Buckheit RW (1991) Inhibition of HIV-LTRgene expression by oligonucleotides targeted to TAR element. Nucleic Acid Res 19:3359–3368

Wagner RW (1994) Gene inhibition using antisense oligonucleotides. Nature 372:333–335

Wagner RW, Matteucci MD, Lewis JG, Gutierrez AJ, Moulds C, Froehler BC (1993) Antisense gene inhibition by oligonucleotides containing C-5 propyne pyrimidines. Science 250:1510–1513

Wahlestedt C, Golanov E, Yamamoto S, Yee F, Ericson H (1993) Antisense oligonucleotides to NMDA-R1 receptor channel protect cortical neurons from excitoxicity and reduce focal ischemic infarctions. Nature 363:260–263

Wallach D (1997) Cell death induction by TNF: a matter of self-control. Trends Biochem Sci 22:107–109

Watson JD, Crick FHC (1953) Molecular structure of nucleic acids. Nature 171:737–738

Watt R, Stanton LW, Marcu KB, Gallo RC, Croce CM, Rovera G (1983) Nucleotide sequence of cloned cDNA of human c-myc oncogene. Nature 303:725–728

Weinberg RA (1996) How cancer arises. Sci Am 62–70

Wickstrom E (1986) Oligonucleotide degradation by nuclease. J Biochem Biophys Methods 13:97–102

Wyatt JR, Vickers TA, Roberson JL, Buckheit RW, Klimkait T (1994) Combinatorially selected guanosine quartet structure is a potent inhibitor of human immunodeficiency virus envelope mediated cell fusion. Proc Natl Acad Sci 91:1356–1360

Yuan J, Horvitz HR (1990) The caenorhabditis elegans genes ced-3 and c3d-4 act autonomously to cause programmed cell death. Dev Biol 138:33–39

Yung WK (1994) New approaches in brain tumor therapy using gene transfer and antisense oligonucleotides. Curr Opin Oncol 6(3):235–239

Zamecnik PC, Stephenson ML (1978) Inhibition of Rous sarcoma virus replication and cell transformation by a specific oligodeoxynucleotide. Proc Natl Acad Sci USA 75:280–285

Zhou L-W, Zhang SP, Qin ZH, Weiss B (1994) In vivo administration of an oligonucleotide antisense to the D_2 dopamine receptor messenger RNA inhibits D_2 dopamine receptor-mediated behaviour and expression of D_2 dopamine receptors in mouse striatum. J Pharmacol Exp Ther 268:1015–1023

Subject Index

acute respiratory distress syndrome 122
Alzheimer's disease 16, 20, 51
amino-pyrimidine 153
amplified mRNA 203
antisense mechanism 229, 253
apoptosis 54
aptamer 136, 142, 146, 151
arabinogalactan 127
arthritis 122
asialoglycoprotein receptor 126
asialoglycoprotein receptor imaging 127
astrocyte 56
astrocyte osmoregulation 56
ATPase flux 57

β-emitting radionuclides 227
brain spectrum 46
breast cancer 19

C-reactive protein 119
carbohydrate-deficient glycoprotein syndrome 124
carbohydrate-deficient transferrin 124
carbohydrates 104, 108, 119, 126
cell adhesion 117, 119, 126
cell adhesion molecules 107
cho 50
choline 54
cocaine abuse 98
combinatorial chemistry 249
combinatorial synthesis 138
congenital dyserythropoietic anemia type II 124
conjugation 227, 230, 253
contrast media 8
counterselection 146
creatine kinase 44
creatine kinase equilibrium 56
Crohn's disease 124

D-galactose 105, 108
D-glucose 105
D-mannose 109, 127
decoupling 58
diagnosis 44
disease processes 16
DNA 142, 226
dolichylphosphate 111
dopamine transporter 97, 98
double-stranded DNA 142
Downs syndrome 62
drug development 95, 99

E-selectin 120, 126
endoplasmic reticulum 111, 116
endothelial cells 126
endothelium 120
enzyme activities 44
epithelial cells 108

epithelial surfaces 104
Europium 62

FIAU 171
fluorinated drugs 62
fluorine-18 deoxyglucose 20
fluoro 152
fucoidin 123
fucosyltransferase 114

galactose receptors 119
galactosyltransferase 124
gangliosides 107
gene delivery 126
gene expression 162, 163, 171
gene therapy 21
gene transduction 163
genetics 15, 20
glioma 170
glutamate 47
glutaminase 59
glutamine 47
glycolipids 107, 111
glycoproteins 108, 111, 114
glycosaminoglycans 107
glycosidases 111, 114
glycosylation 114
glycosylphosphatidylinositol (GPI) membrane anchors 107
glycosyltransferase 111, 124
Golgi apparatus 111, 114
GPC 50
GPE 50
GPI anchors 108

heart function 24
heparan sulfate 107
heparin 107
hepatic encephalopathy 56
hepatocytes 119, 126
herpes simplex virus 163
homeostasis 17
hybridization 225, 227, 230, 250
hypoxia 48

imaging 43, 53, 68, 70, 163, 176
in vitro selection 136, 142, 147
in vivo magnetic resonance spectroscopy 43
inflammation 119
inorganic phosphate 57
inositol phosphates 62
interaction 202, 230, 236, 249
International Classification of Diseases 16

Kupffer cells 119, 126

L-fucose 105, 108, 127
L-selectin 120
lactate 48
lectins 119, 126
leukocyte adhesion deficiency II 124
leukocytes 119, 126
liver 126
liver tumors 127
lymphe node 120
lysosomal storage disorders 124
lysosomes 119

macrophages 119, 126
magnetic resonance 23, 33
magnetic resonance imaging 33, 36
malignancy 104
mannose 119
MAP kinase 55
marrow transplant 62
medical imaging 1
meningoencephalitis 122
metabolic disorders 21
metabolite 44
metabolite concentration 44, 48, 63

metastasis 126
molecular imaging 84
molecular recognition 137
monoclonal antibody 68, 74
MR scanners 48
mRNA 202, 225, 238, 247
mucins 104
multidrug resistance 18
multinuclear magnetic resonance spectoscopy 51

N-acetyl-D-galactosamine 105, 111
neurones 56
neuroreceptors 92, 96, 99
nuclear magnetic resonance 11
nuclear medicine 9
nucleic acids 138, 146

O-methyl 152
oligonucleotides 202, 227, 249, 251
oligosaccharide processing 112
oligosaccharides 108, 111, 112, 119

p-Glycoprotein 18
P-selectin 120
Parkinson's disease 92, 98
paroxysmal nocturnal hemoglobinuria 124
PC 50
PE 50
peptides 67, 73
perfusion 34, 39
PET 166
platelets 119
polymerase chain reaction 142
polysaccharide 105, 107
positron emission tomography 53, 83, 98
positron-emitting tracers 83
protein folding 116
proteoglycans 107, 111

quantitation 48

radioactive tracers 16, 18
radioimmunoscintigraphy 68, 72
radiolabeling 203, 225, 227
random sequences 142
receptor 44, 53, 107, 119
reperfusion injury 122
reverse transcription 144
rheumatoid arthritis 124
ribozymes 141
RNA 136, 142, 152
Roentgen, Wilhelm Conrad 2

selectins 119
SELEX 142, 144, 151
sensitivity 85
sialic acids 105, 108
sialyl Lewis A 120, 126
sialyl Lewis X 126
signaling pathway 202, 223, 236, 241, 246
single-stranded DNA 142
single-stranded nucleic acids 140
"smart" contrast agents 62
SPECT 166
spiegelmer 153
stability 150
stress 56
stroke 35, 39
substrate chase 59
superfamily 240, 243
superparamagnetic iron oxide 127

target 136, 144
target tissue 150
targeting 135
technetium-99m (^{99m}Tc) Sestamibi 18

testosterone 144
therapeutic monitoring 44
thymidine kinase 21, 163
transgenic 62
translation 84
tRNA 140
tuberculosis 124

vascular epithelial growth factor 136

X-rays 3

Ernst Schering Research Foundation Workshop

Editors: Günter Stock
Ursula-F. Habenicht

Vol. 1 *(1991)*
Bioscience ⇌ Society
Workshop Report
Editors: D. J. Roy, B. E. Wynne, R. W. Old

Vol. 2 (1991)
Round Table Discussion on Bioscience ⇌ Society
Editor: J. J. Cherfas

Vol. 3 (1991)
Excitatory Amino Acids and Second Messenger Systems
Editors: V. I. Teichberg, L. Turski

Vol. 4 (1992)
Spermatogenesis – Fertilization – Contraception
Editors: E. Nieschlag, U.-F. Habenicht

Vol. 5 (1992)
Sex Steroids and the Cardiovascular System
Editors: P. Ramwell, G. Rubanyi, E. Schillinger

Vol. 6 (1993)
Transgenic Animals as Model Systems for Human Diseases
Editors: E. F. Wagner, F. Theuring

Vol. 7 (1993)
Basic Mechanisms Controlling Term and Preterm Birth
Editors: K. Chwalisz, R. E. Garfield

Vol. 8 (1994)
Health Care 2010
Editors: C. Bezold, K. Knabner

Vol. 9 (1994)
Sex Steroids and Bone
Editors: R. Ziegler, J. Pfeilschifter, M. Bräutigam

Vol. 10 (1994)
Nongenotoxic Carcinogenesis
Editors: A. Cockburn, L. Smith

Vol. 11 (1994)
Cell Culture in Pharmaceutical Research
Editors: N. E. Fusenig, H. Graf

Vol. 12 (1994)
Interactions Between Adjuvants, Agrochemical and Target Organisms
Editors: P. J. Holloway, R. T. Rees, D. Stock

Vol. 13 (1994)
Assessment of the Use of Single Cytochrome P450 Enzymes in Drug Research
Editors: M. R. Waterman, M. Hildebrand

Vol. 14 (1995)
Apoptosis in Hormone-Dependent Cancers
Editors: M. Tenniswood, H. Michna

Vol. 15 (1995)
Computer Aided Drug Design in Industrial Research
Editors: E. C. Herrmann, R. Franke

Vol. 16 (1995)
Organ-Selective Actions of Steroid Hormones
Editors: D. T. Baird, G. Schütz, R. Krattenmacher

Vol. 17 (1996)
Alzheimer's Disease
Editors: J.D. Turner, K. Beyreuther, F. Theuring

Vol. 18 (1997)
The Endometrium as a Target for Contraception
Editors: H.M. Beier, M.J.K. Harper, K. Chwalisz

Vol. 19 (1997)
EGF Receptor in Tumor Growth and Progression
Editors: R. B. Lichtner, R. N. Harkins

Vol. 20 (1997)
Cellular Therapy
Editors: H. Wekerle, H. Graf, J.D. Turner

Vol. 21 (1997)
Nitric Oxide, Cytochromes P 450, and Sexual Steroid Hormones
Editors: J.R. Lancaster, J.F. Parkinson

Vol. 22 (1997)
Impact of Molecular Biology and New Technical Developments in Diagnostic Imaging
Editors: W. Semmler, M. Schwaiger

Supplement 1 (1994)
Molecular and Cellular Endocrinology of the Testis
Editors: G. Verhoeven, U.-F. Habenicht

Supplement 2 (1996)
Signal Transduction in Testicular Cells
Editors: V. Hansson, F. O. Levy, K. Taskén

MIX
Papier aus verantwortungsvollen Quellen
Paper from responsible sources
FSC® C105338

If you have any concerns about our products,
you can contact us on
ProductSafety@springernature.com

In case Publisher is established outside the EU,
the EU authorized representative is:
Springer Nature Customer Service Center GmbH
Europaplatz 3, 69115 Heidelberg, Germany

Printed by Libri Plureos GmbH
in Hamburg, Germany